THE Kingdom of God IS **GREEN**

THE Kingdom OF God IS GREEN

Paul Gilk

WIPF & STOCK · Eugene, Oregon

THE KINGDOM OF GOD IS GREEN

Copyright © 2012 Paul Gilk. All rights reserved. Except for brief quotations in critical publications or reviews, no part of this book may be reproduced in any manner without prior written permission from the publisher. Write: Permissions, Wipf and Stock Publishers, 199 W. 8th Ave., Suite 3, Eugene, OR 97401.

Wipf & Stock
An Imprint of Wipf and Stock Publishers
199 W. 8th Ave., Suite 3
Eugene, OR 97401
www.wipfandstock.com

ISBN 13: 978-1-61097-537-7
Manufactured in the U.S.A.

For Nils Pearson, who gave me the right book at the right time, and for Carol Ann Okite, who deserves far more from me than a passing reference on the Acknowledgments page—my love to you both.

Contents

Foreword | ix
Acknowledgments | xi
Introduction | xiii

BOOK ONE

1 A Child's View of Progress | 3
2 The Gorgeous Chalice of Civilizational Myth | 5
3 My Friend Pagus | 8
4 Holding the Hand of Conservative | 11
5 Sacred Cow Crap | 14
6 A Left-Wing Odd Duck | 22
7 What Is "Religionless Christianity"? | 25
8 The Aristocracy of Consumption | 29
9 The Kingdom of God Is Green | 36
10 An Ethos Spawned and Nourished | 43
11 Servanthood, Stewardship, and a Restraint on Vice | 52
12 Clinging to Dead Ideas | 61
13 The Engine of Disaster | 71
14 Voting for Jesus | 78
15 The Gathering Globalization of Disaster | 96
16 Packed in a Comforting Mythology | 101
17 The Superlative Proportions of Our Self-Inflation | 106
18 Two Losers and an Icon | 111
19 Ending the Bogeyman Cycle | 127
20 The End of Something Big | 135
21 Divine Terrorists | 141
22 The Gendered Feminine and Twenty Centuries of Papa | 142

BOOK TWO

23 Age of the Daughter | 145
24 Stuff | 194
25 A Broken Bone | 198
26 An Admonition to Gore Vidal | 201
27 The Imposition of Those Glories | 208
28 The Village of God | 220
29 This Predator Beast Game | 227
30 The Feminine Dimensions of God | 235
31 Such Pureness of Heart | 238
32 God's Lifeboat | 243
33 The Global Cloning of Civilized Desire | 253
34 The Gardener from Amenia | 256
35 Pushing One Hundred | 260
36 One Last Thing | 264
37 Civilization, "Civilization," and the Kingdom of God: An Afterword | 267

A Note on Source Material | 272
Bibliography | 275

Foreword

THE FOLLOWING ESSAYS MAY strike some people—maybe most people—as perverse. I have good friends who don't like what I say about civilization or orthodox religion. Leaving aside for the time being the subject of religion (there's plenty of that in the following essays), I will only say here that I came by my conclusions regarding civilization the hard way: I looked up at its history from the earthy perspective of the lowly peasant.

When I finally got around to reading Howard Zinn's deservedly famous book *A People's History of the United States*, I'd already come to the conclusion he articulates on page 8 of his opening chapter: "The historian's distortion is more than technical, it is ideological." I am now inclined, however, to say "mythological" rather than "ideological." Perhaps it's both mythological and ideological. Let's just say it's both.

As to the core of my conviction—clearly not only my conviction—allow me to quote from the massive work of William H. McNeill, *The Rise of the West*, page 38. This is McNeill's section where he talks about the transformation of agrarian villages ("remarkably peaceful societies") into the rudiments of aggressive civilization "when pastoral conquest superimposed upon peaceful villagers the elements of warlike organization from which civilized political institutions without exception descend."

For McNeill, herders ("pastoralists") were closer to hunting warriors and they therefore tended to congeal in patrilineal families under the authority of a chieftain. McNeill suggests it was the raiding aggression of such men, with their subsequent permanent takeover of the agrarian villages, from which *civilized political institutions without exception descend*. (On page 35, McNeill also suggests that matrilineal family systems prevailed in the agrarian villages along with female deities, but that "male deities date only from later times, when political and other changes had drastically altered the social structure of the earliest Neolithic village communities.")

All this is why I say our mass reverence for "civilized values" is of such magnitude that it spills way beyond the mold of ideology and on

Foreword

into the mold of mythology. God and civilization are the pre-eminent cosmic magnets aligning our political souls. In these essays we will delve into the dangerous dynamics in this magnetic field.

Acknowledgments

IT'S TRUE THAT I spend a lot of time at home, in the log house, in the woods. But everybody needs a community, a network, and much of mine consists of our small Quaker Meeting, a stubbornly persistent Peace Study, the Holy Cross sisters (whose American headquarters is in Merrill, with the international in Switzerland), and the wonderfully helpful librarians at the beautiful T. B. Scott Public Library, with its big windows overlooking the Prairie River, also in Merrill.

There's no way to adequately say how much these folks have helped to keep me intellectually alive. But I thank them one and all.

Introduction

MOST OF THESE ESSAYS were composed in the years immediately after 9/11—a numerical designation that almost instantly became iconic and, for those with a somewhat opaque political agenda, a designation that also became a kind of numerological occultism or talismanic charm. I say this with a touch of sarcasm, perhaps—"numerological occultism" and "talismanic charm" have a bit of biting whimsy in them—but the sardonic terms also point with genuine alarm to the raw outburst of empire hubris so wantonly deployed by the George W. Bush administration in the aftermath of 9/11. Dick Cheney's sneer may be the image that truly lingers.

Yet these essays are not built primarily around 9/11 or the politics and policies enabled by 9/11, even though most were written in the shadow of 9/11. (To some extent I felt and still feel that the shock of 9/11 generated an involuntary but wholesome deepening of analysis on the part of various writers and intellectuals on the Left. Shocks can—and will—cause such deepening. To a degree, my essays are reflective of that deepened energy, an energy "discovered" as shock broke through layers of conventional conviction and comfortable behavior to liberate deeper emotions and stimulate richer analysis.)

But what oriented me intellectually was already taking shape in the early 1970s. Although I was then an alienated young man living in the city of St. Louis, the question that proved an abiding burr under my saddle was—"Why are small farms dying?"

I'd grown up on a small homestead farm in northern Wisconsin and, in inner-city St. Louis, a long way from home, I discovered that I not only missed the overall atmosphere of rural life but, aware that small farms were dying, I wanted to know *why* they were dying. Getting only trite and conventional answers from those I thought smart and informed, I found myself increasingly determined to get to the bottom of this barrel. I soon began to realize that almost everyone—if they even bothered to think about it—had only superficial and pat answers

Introduction

to questions about agriculture's relationship to civilization. Part of my learning curve, one might say, was coming to terms with this—pardon my English—mass stupidity regarding our past, especially our agrarian past.

What I initially thought a fairly narrow but deep avenue of inquiry began to open like a cleverly folded oriental fan, revealing a panorama of related issues whose complexity was stunning. By looking into the origins of agriculture I discovered an ancient history of ancestors stretching way back to our primate cousins, the multimillion-year evolution of our species, the hugely long time we spent in caves with firelight, the long cultural gestation our ancestors were in as gatherers and hunters with Stone Age tools, the break-out development of gatherers creating horticulture after the last Ice Age, the subsequent domestication of animals who could also eat from this horticultural abundance, the establishment of settled villages—and then, suddenly, the scaffold of civilization, the suppression of a feminine spirituality linked (in part) to horticulture, the imposition of systemic slavery and permanent militarism, the building of astonishing empires bristling with aggression, the erection of a male-forged utopian psychology (wandering back and forth between utopian religion and utopian metaphysics) that began to become universal with Euro-American globalization, the industrial revolution, and the liquidation of what we might call indigenous or folk sensibility. (The reader can find most of my previous book titles in the Bibliography, as well as works by Lewis Mumford and Norman O. Brown that were especially foundational to my thought.)

I'd been raised a farm boy who, with his family, went regularly to church—the United Church of Christ, to be specific. By age seventeen I was, one might say, cooked. I'd had it. The religious stuff was a weird blend of induced otherworldly preoccupation and metaphysical abstraction (over the altar at St. Stephens was a huge painting of Jesus ascending into the sky, to the utter amazement of disciples gazing up rapt from below). These otherwordly expectations and gravity-defying abstractions never connected with my actual life, despite a serious shot of will power and so-called positive thinking. That's not to say that Sunday school Bible stores hadn't made an impact. They had and they did. But I literally spent decades—sometimes with bewildered focus, sometimes with bitter flaccidity—trying to clarify what I believed and attempting to sort out what was true—or, at least, what was true for me.

Introduction

Beginning about 1990, I resolved to read a chapter a day from one or another of the four canonical Gospels. I did this, virtually without a break, for about fifteen years. It took most of a decade for me to realize how central the "kingdom of God" proclamation was—and is—in the first three Gospels. And just as this awareness was breaking in on me, a Quaker friend—Nils Pearson—introduced me to the writing of Marcus Borg and, through Borg's allusions and references, to John Dominic Crossan.

With ten years, more or less, of Gospel reading under my belt, topped by the scrupulous (and wonderfully readable) scholarship of Borg and Crossan in particular, I began to understand the nature of civilized authority in a more comprehensive way. When Borg methodically unfolded the implications of the monarchical model (or image) of God in *The God We Never Knew*, I began to understand far more deeply how the notion of an all-powerful, elsewhere God works like a psychological and political magnet to align obedience and conformity in human behavior, including long-term obedience and multigenerational conformity to systems of perpetual exploitation. Combine this insight with the literally immeasurable influence of Augustine's doctrine that it's God who gives and sustains kingdoms and empires (that is, God is the divine energy behind and upholding civilization), and it becomes possible to discern how God-image correlates to political structure, what it meant for "Jesus" to wed his murderer—the Roman Empire—in the fourth century.

And that—how God-image correlates to power and to the ways human life on Earth is organized by civilized power—is one way of describing the subject or theme of these essays.

Of course, the conventional or dominant image of God is as cosmic Father. That is, God is not only the most powerful Being in (or even beyond) the universe, God is also Male. This masculine depiction of God correlates historically to all three Abrahamic religions—Judaism, Christianity, and Islam—and it's by no means incidental that all three religions arose in a historical and geographical context in which the overarching established power was civilizational. Male rule—the monarchical model—was the cultural norm. Therefore God is King of kings. But civilization globalized paradoxically magnifies male rule to ecocidal dimensions even as the from-below energy of the marginalized and dispossessed forces onto civilization an ongoing, unresolved ferment of economic restraint, equality, and democracy; and that includes equality for women and, with it, a comprehensive feminist critique and a spiritually infiltrating feminine ambience of the sacred.

So, for Christians, God has been depicted as totally male and elsewhere (though always listening in), and the real life to come (i.e., the life after death) will also be elsewhere. God will burst in—not from below—but from above when He decides it's time to bring Earthlife to a close and separate the (relatively few) sheep from the (multitudinous) goats. Civilization, correspondingly, has been aristocratic and utopian (that is, psychologically gated and otherworldly) from its inception: see, please, Mumford's magnificent essay "Utopia, the City, and the Machine" in his *Interpretations and Forecasts*. Where God and Civilization most powerfully congeal is in the otherworldly Christian unity forged by Constantine and rationalized by Augustine. The vicious, enduring epithet that reveals their pathological contempt for earthly nature and rural life was and is *paganism*.

It took me a while to do so, but I eventually discovered that the mythology of civilizational origins (civilization supposedly growing "naturally" out of agrarian abundance) and the mythology of religious creationism (agriculture as God's curse on humanity for Eve and Adam mucking it up in Eden) reinforce each other by obstructing a truer and more wholesome understanding of cultural origins and cultural evolution. Christian fundamentalism—which is nothing more than traditional Christian orthodoxy in a contemporary posture of stripped-down, aggressive defensiveness—simply denies that there was life before the Genesis account of Creation. Civilizational mythology, because it has fairly strong threads of rationality in its political make-up (it's impossible to keep empire functioning without ruthlessly calculating *Realpolitik*, and that necessitates a certain kind of focused rationality*)*, can acknowledge "prehistory" but, as a rule, it treats that knowledge with dismissive condescension and aristocratic contempt. "Paganism" as a deadly epithet suits both the civilized elite and the Christian orthodox because it combines in deadly concentration civilized contempt for the noncivilized and religious dread of all those disinclined to enter the psychologically totalitarian City of God. If *The City of God* represents the glowing, positive image of Constantinian Christianity, *paganism* is its Manichean depiction of rural evil—the Hobbesian chaos that civilized order will contain and control and, if necessary, crush.

We can therefore say (this Introduction is a severe compaction) that civilization's globalization occurs simultaneously with the destruction

Introduction

of the indigenous and the liquidation of the peasant. Insofar as God and Civilization are two utopian peas in an otherworldly, metaphysical pod, we can also say that the ecological landscape of planet Earth is not only under threat from utopian economic obliviousness but that our political landscape remains largely the magnet field of authoritarian utopianism—policies and behaviors so powerfully ungrounded in an earthly sense that the utopian construction of global apocalypse is deemed normal—as, for instance, in the manic insistence on a constantly growing industrial economy even in the face of planetary climate change that will become increasingly (and perhaps irreversibly) disastrous.

In a broad and sweeping way, we might say the Republican Party is composed almost entirely of either ideological utopians or Augustinian fundamentalists or some shifting amalgam of the two. The Democratic Party—certainly at its upper policy-making end—is composed of secular (but also ideological) utopians. At its lower level there are some alarmed eutopian radicals who recognize the immediate and pressing need for (among other things) a huge reduction in fossil fuel consumption and carbon emissions. In-between the secular utopians and eutopian radicals there is (I think) a majority who are sluggishly waking up from their affluent, utopian somnolence, but who are still too spiritually numb and historically novocained to do much more than scratch and stretch and yawn. They're not dumb; they just haven't spent nearly enough time wrestling with the issues. Utopia has simply been too encompassing, engrossing, comfortable, and ubiquitous. The Empire cafe, however, is about to have its menu massively downsized.

To achieve an ecological and humane *eutopian* outcome to the rapidly escalating crises requires the abandonment of God and Civilization as governing constructs and controlling concepts. Instead, we need the embrace of *Spirit*, with restoration of the folk evolution that was culturally stoppered when civilized aristocrats (i.e., feral male bandits and brigands) forced peaceful and modestly well-off agrarian villagers into becoming desperate, destitute peasants. The "kingdom of God" proclaimed by Jesus implies the resurrection of the agrarian village, free of civilized oppression. The feral males who continue to own and operate corporate civilization need the domesticating and gentling embrace of the feminine village.

It finally dawned on me that the "kingdom of God," despite its ridiculous semantic baggage, is exactly about this embrace and resurrection. There is no question that Jesus taught what we might call radical egalitarian servanthood. Perhaps our contemporary ecological crisis raises ecological stewardship to an ethical position on a par with egalitarian servanthood. Well, yes and no. Maybe not really on a par. We humans are only a small part of Creation. Despite our manic self-inflation and stunningly narcissistic hubris, we are still only one species on Earth—even as we seem contemptuously determined to whittle away at the "competitors."

Servanthood is really a derivative ethical understanding emerging from our evolution on this awesome Earth. If Earth—or the Spirit that animates life and Earth—could nurture us through all those countless millennia and not snuff us out, then compassionate servanthood really seems like a beautiful idea to lavishly practice and gratefully pass on. It is, after all, what's enabled us to get here.

The "kingdom of God," we might nevertheless say, stands on two ethical legs: radical servanthood and radical stewardship. We have to treat each other and all living things in a far more loving and reverent way. And, since advocates of Green culture and Green politics have been explicitly promoting ecological stewardship and egalitarian servanthood for decades, we can therefore without hesitation assert that the "kingdom of God" is Green.

BOOK ONE

We are beginning a mythic period of existence, rather like the age portrayed in the *Bhagavad Gita*, in *The Lord of the Rings*, and in other tales of darkness and light. We live in a time in which every living system is in decline, and the rate of decline is accelerating as our economy grows. The commercial processes that bring us the kind of lives we supposedly desire are destroying the earth and the life we cherish. Given current corporate practices, not one wildlife reserve, wilderness, or indigenous culture will survive the global market economy. We are losing our forests, coral reefs, topsoil, water, biodiversity, and climatic stability. The land, sea, and air have been functionally transformed from life-supporting systems into repositories for waste.

—PAUL HAWKEN
"The Resurgence of Citizens' Movements,"
in *Utne Reader* (November-December 2000), page 1

In *The Shock Doctrine*, I explore how the right has systematically used crises—real and trumped up—to push through a brutal ideological agenda designed not to solve the problems that created the crises but rather to enrich elites. As the climate crisis begins to bite, it will be no exception. This is entirely predictable. Finding new ways to privatize the commons and to profit from disaster are what our current system is built to do. The process is already well under way.

The only wild card is whether some countervailing popular movement will step up to provide a viable alternative to this grim future. That means not just an alternative set of policy proposals but an alternative worldview to rival the one at the heart of the ecological crisis—this time, embedded in interdependence rather than hyper-individualism, reciprocity rather than dominance and cooperation rather than hierarchy.

—NAOMI KLEIN,
"Capitalism vs. the Climate,"
in *The Nation* (November 28, 2011), page 19

To be logically consistent the environmentalist must have a basically religious commitment to sustainability. There is no other answer to the logical naturalist who says that the sustainability of human society is nothing but an anthropocentric conceit.

—HERMAN E. DALY,
"Economics and Sustainability: In Defense of a Steady-State Economy,"
in *Deep Ecology*, page 97

To doubt that Spirit infuses Nature is to doubt that citizenship is possible.
—J. RONALD ENGEL,
"Liberal Democracy and the Fate of the Earth,"
in *Spirit and Nature*, pages 78–79

1

A Child's View of Progress

CORRECT ME IF I'M wrong, but I was taught that civilization always and everywhere was made up of aristocrats and peasants until the industrial revolution was invented. Then new machinery made human slavery unnecessary. And that apparently means that before the industrial revolution, slavery *was* necessary. That's pretty scary, but I guess I understand. And then—let me see if I've got this right—human beings were able to democratize civilization because slavery could now be imposed directly onto nature, because of machinery. Mechanization, if I'm on the right track, enabled an absolutely huge increase in energy consumption (to power the machines) and commodity production (the things machines made), so that virtually everybody could be just awash in stuff. And, not being serfs or slaves any longer, we were free to be Democrats or Republicans, "liberals" or "conservatives." We could vote for our rulers. So we have elections now and we're done with slavery, done with aristocracy. We've got stuff, and we are living in the promised land of civility, thanks to the mechanical enslavement of nature. We have Taylorism, robotics, genetic engineering, agribusiness, plenty of weapons to protect all this stuff, and we're as happy as pigs in mud, although I apologize for the vulgar image.

Oh, we have a few wrinkles to iron out. According to the newspapers and magazines there are such things as child poverty, a prison *industry* (though I can't quite figure out what they make there), sweatshop labor, vertical integration in agribusiness, massive corporate monopolies, species extinctions, genetic mutations, depletion of the oceans' fish stocks and destruction of the jungles' rain forests, plagues of new diseases, a lot of chemical toxins, radioactive waste, apocalyptic weapons, climate change, and global warming—but, hey! Like my smart teacher puts it, who said anything about a free lunch? Technical fixes. That's the term to remember. That's what will enable this nature-slave system to keep right on a-rolling. Because it's *civilization*, stupid, and civilization is good. It's what got us here. It's what everyone aspires to.

See, let me explain this once more, just to see if I've got it right. Once upon a time civilization was maybe a little bad because aristocrats made peasants do all the work, and aristocrats took away most of what the peasants produced or made. See, there just wasn't enough stuff for everybody to get rich. So only aristocrats could get rich. They were a kind of chosen people—maybe like winners of a lottery. But you had to have a sword to enter, and not everybody could afford one.

And then along came the industrial revolution, see, and the people who owned those machines made the machines do what the aristocrats used to make the peasants and workers do. So now everybody could be just like the aristocrats. Everybody could get rich. Everybody could have lots of stuff. See? Because we all wanted so badly to be civilized. Because we all love civilization.

And so civilization really is a slavery system, see, but now we've gotten so smart and so good that we just enslave nature. See? Nature doesn't mind at all. Nature likes to be enslaved. My economics teacher told me so. And so did my civics teacher. Correct me if I'm wrong, OK?

2

The Gorgeous Chalice of Civilizational Myth

IT SEEMS A REASONABLY safe thing to say that the extending and deepening of civilized forms of behavior is what's dominant in the modern world. I mean, in particular, civilized behaviors shaped by the great burst of world-conquering energy that erupted out of Europe in the fifteenth through the seventeenth centuries, and then the subsequent industrialization of that energy, patterned after England's industrial revolution. Industrial progress became the byword of human aspiration. Also known as "development," progress was bringing civilization and civilized values to uncivilized peoples and underdeveloped regions around the globe.

Christian missions fit right in with this effort to lift the dark world of heathenism and paganism out of darkness and on into the bright, clear, clean world of Christian civilization. Humanity as a whole was finally getting a grip on its backwardness, finding traction to lift itself out of darkness. The world was becoming a much finer place, and at a truly amazing pace of achievement. Nothing like this had ever happened before in human history—at least not on this scale and with this degree of inclusion. The sheer magnitude of its totality was awesome. Not only was there light at the end of the tunnel, the human race as a whole was in process of climbing out of the dark tunnel of primitivity and on into the glorious sunlight of a bright, new word of civility. Now everyone could be civilized.

But something odd has happened on the road to utopian perfection. We are, in fact, in the midst of unprecedented crises, of traumas and tumults clearly associated with civilized globalization: weapons of mass destruction, climate change, species extinctions, ecological disasters of various kinds, human overpopulation. How did this energetic, exceedingly optimistic project of perfection produce such toxic fruits?

There is a very interesting Christian writer and scholar by the name of Gil Bailie. He has a book entitled *Violence Unveiled: Humanity at the Crossroads*. One of the book's themes is the role myth plays in religion, culture, and everyday human psychology. Myth, says Bailie, comes from a Greek word, *muthos*, "which means 'to close' or 'keep secret.'" *Muthos*, says Bailie, is related to *muo*, which "means to close one's eyes or mouth, to mute the voice, or to remain mute."[1]

Now, of course, we all were taught that civilization is unconditionally a good thing, absolutely so. Yet if one reads books on the origins of civilization—say Lewis Mumford's *The Myth of the Machine* and *The Pentagon of Power* or Arnold Toynbee's *Civilization on Trial*—what one learns is that civilization came into being with "traumatic institutions" (Mumford) and "diseases" of Class and War (Toynbee). Myth, meanwhile, has beautifully covered over and elegantly disguised the brutal violence by which vicious bandits imposed a regime of permanent expropriation on agrarian villages and then used that extorted "surplus" to build the first cities governed by aristocrats. Civilization is constructed by violence, extortion, and expropriation.

So civilization has come down to us as a thing of shining beauty and moral perfection when, in fact, its core is ruthless violence and rapacious greed. So strong is this myth of lofty civility, so deep and pervasive is it in human consciousness, so reflexively taught as supremely true, that to disbelieve it (much less to deconstruct it) is to put the disbeliever in a vulnerable zone between heresy and treason. So we not only have a pattern of incredibly penetrating indoctrination, we also have an indoctrination nearly impossible to break out of because of the *moral* grip the myth has on us—an evil thing dressed in the awesome garb of radiant light. We are both captivated by the light and (perhaps unconsciously) terrified by the underlying evil with its power of condemnation and wrath.

Gil Bailie goes on to say that in the New Testament, "*mythos* is juxtaposed to *Logos*—the revelation of that about which myth refuses to speak—and to *aletheia*—the Greek word for truth. . . . The literal meaning, then, of the Greek word for truth, *aletheia*, is 'to stop forgetting.' It is etymologically the opposite of myth."[2] In other words, to see the multimillennial project of civilization for what, at its core, it truly is requires deep immersion in *aletheia*, in truth, in the spiritual and intellectual courage to stop forgetting and to speak the truth out loud.[1]

But all this is tantamount to calling civilization a multimillennial project of war criminality with an underbelly of economic extortion and ecological oblivion. This is, needless to say, a view that continues to be incredibly difficult to accept. Yet the accelerating and intensifying crises are forcing us to face this fact, this myth-protected monster. The only question is—What magnitude of disaster does it take to shatter this gorgeous chalice of civilizational myth?

Notes

1. Bailie, *Violence*, 33.
2. Bailie, *Violence*, 33.

I. Brian D. McLaren's *Everything Must Change: When the World's Biggest Problems and Jesus' Good News Collide* is a book glimmering with brilliant insights. McLaren doesn't seem inclined to keep secrets or close his eyes. But in my estimation he's not looking deeply enough. Two examples spring to mind.

On page 5, he says that contemporary global crises are symptoms "of a *suicide machine* that co-opts the main mechanisms of our civilization—our economic, political, and military systems—and reprograms them to destroy those they should save." On page 83, in talking about the orthodox or conventional view of Jesus and its supposedly unintended negative consequences, he says "our conventional view has accidentally put Jesus in the very framing story Jesus originally sought to subvert."

But what if the "suicide machine" is *not* "co-opting" the main mechanisms of our civilization but is, instead, bringing those mechanisms to global triumph? That is, what if the crises are the logical outcome of the "main mechanisms"?

Somewhat similarly, we might refrain from impulsively excusing the framers of the conventional Jesus as "accidentally" putting Jesus into an otherworldly context. Those framers did so deliberately. To say otherwise is historically false. No doubt they could not have foreseen the long-term consequences of the Constantinian Arrangement, but that doesn't obviate their intentionality.

Perhaps "smoke screen"—not necessarily created out of cynicism, but the product of well-intended naïveté—is a halfway position between *mythos* and *aletheia*. But if I understand what Bailie means by *aletheia*, it's an imperative that doesn't stop with the overcoming of muteness but continues on to suck the disguising smoke out of our naïveté, as well.

3

My Friend Pagus

THERE WERE A COUPLE things I learned in school. On Sunday, I learned that Jesus is my savior. On Monday, I learned that Civilization also represents my salvation. In the first instance, all I had to do was "believe in" Jesus and a heavenly life, after death, would be given to me. I would be "saved." In the second instance, just as Jesus was saving me from the wages of sin, Civilization had saved me, and was still saving me, from primitivity, paganism, heathenism, and backwardness. Then I learned, back in Sunday school, that Jesus was saving me from some of the same things Civilization was saving me from, especially heathenism and paganism.

It was good to know that I was a civilized Christian, living in a Christian Civilization. I felt safe and saved.

But there were two small flaws in my redemption. One was my curiosity about lots of things, including words—and my curiosity was expanding. The other flaw was growing up on a small farm, and small farms were dying. These flaws collided with my comfortable convictions one day when a plump *Webster's* informed me that *civis*, city, was the Greek root of civilization, and that *pagus*, country place, was the Latin root of peasant. "[See PAGAN]" said the dictionary, in its definition of peasant. So, what the heck, I saw pagan. And, to my astonishment, there was my friend *pagus*—etymological mother of twins! Not only was *pagus* the linguistic womb of *peasant*, that ancient race of exploited and expropriated gardeners, she was also the rural womb of *pagan*, about whom I had been so severely and righteously warned. I was both shocked and bewildered. I had no idea that peasants were by definition pagans.

So I paged through the Gospels. And, sure enough, there was Jesus saying all those *biological* things about vine and branches, about being a good shepherd, about stewards, about seeds and yeast, about bread and wine. I read more closely. Here were terribly fierce things said *against* the rich, *against* those who "lord it over," *against* the rulers, priests, and professors, and *for* the reverential but egalitarian kingdom of God. And then, to cap it all off, the civilized *and* religious leaders arranged the legal assassination of this wandering healer, teacher, and sharer. They killed him dead.

Wait a minute! All kind of bells and buzzers were going off! If Jesus and Civilization were tooting the same tune, how come Civilization killed Jesus?

Well, let's be fair here. In the interest of fuller disclosure, let's just say that by the time I read [see PAGAN] I had already embarked on a determined journey to the roots (if I could find them, if I had the courage to get there) of the "farm crisis."[1]

Here's what I learned—from books, but not in school. Gatherers began to develop horticulture in the way back distant time; let's say ten thousand years ago, to nicely round out the date. Over many centuries this slowly growing horticultural abundance resulted in a number of things. Our ancestors no longer had to be nomadic hunters and gatherers. Villages became stable in location. Human population began to increase. A variety of previously hunted animals were domesticated. Agriculture emerged. It was a time of unprecedented prosperity.

Then something happened. Armed bandits, male "warriors," took over and stole this abundance, this "surplus." They formed an aristocracy, expropriating as much wealth as possible from the agrarian village. This aristocracy of armed men formed an army, enslaved and coerced those whom the army conquered, built cities, and created Civilization. Master and slave, aristocrat and peasant, king and subject: these are the basic human dynamics of all civilizations.

Through a long process of conquering and expansion, Civilization spread slowly over much of Asia, North Africa, and Europe. Beginning in the late fifteenth century, "Christian" Civilization erupted from Europe and spread all over the world, manifesting itself largely through conquest, looting, colonialism, and slavery. With the industrial revolution, Civilization discovered it no longer had use for either peasants or slaves. Machines

---

1. My first book was *Nature's Unruly Mob: Farming and the Crisis in Rural Culture*, published in 1986 by Anvil Press and then (with some revision) republished in 2009 by Wipf and Stock, with a Foreword by Helena Norberg-Hodge, author of *Ancient Futures: Learning from Ladakh*.

worked better. A little over two centuries into the industrializing process (if one dates it from the 1760s, in England), civilized agribusiness has thoroughly crucified the peasants. The good shepherd. The sower of seeds. Stewardship and sufficiency. All gone. Killed them dead.

I use the word "crucified" with firm intent. From the point of view of the peasant, Civilization is the name of a process, a procedure, an aristocratic organization, based on organized theft and institutionalized violence. From the point of view of the peasant, the peon, the serf, and the slave, Civilization is a world-class criminal that has yet to be brought to justice, either human or divine.

With the industrial revolution, Civilization began to sweep the peasant, peon, serf, and slave off the table of history and bury them in an unmarked mass grave called "primitivity." We might call it a case of cultural genocide. In their place, Civilization invented chemicals, machines, and terminator seeds, and calls it "development."

So "Civilization" is: warrior energy congealed in traumatic institutions of violence and theft; contemptuous toward sufficiency, simplicity, stewardship, and sharing; and a world-class criminal masquerading as our savior, quite literally the Anti-Christ.

Just a little footnote to what's *not* taught in school, either on Sunday or on Monday.

4

Holding the Hand of Conservative

THE TWELFTH CHAPTER IN Joan Chittister's *Wisdom Distilled from the Daily: Living the Rule of St. Benedict Today* is entitled "Stability: Revelation of the Many Faces of God." Its theme is—no surprise—"the spirituality of stability."[1]

Stability is a word that suggests a regularity of life, a pattern slow to change—and change only when needed. Many of us have an immediate affection for "stability" and, at the same time, an uneasy suspicion toward a word that should be in the same camp—namely "conservative." What I'd like to do here, briefly, is look at the prevailing use of the word "conservative" in light of a grounded stability.

First some definitions. Stability comes from the Latin *stabilas*, meaning steadfast, steady, and enduring. Conservative comes from conserve, whose Latin is *conservare*, to keep or preserve in a safe or entire condition. The word "conservation" is, obviously, close to the Latin root. But, with the possible exception, for instance, of the mainstream "conservative" position on abortion, the current political use of the term "conservative" is the antithesis of *conservare*. No group is more in favor of unimpeded industrial growth, the breaking down of trade barriers, the jettisoning of workplace regulation, the elimination of the social safety net, and the repealing of environmental protection than—"conservatives."

I am, at the moment, less interested in *how* this doublespeak has happened than that it *has* happened. And, furthermore, I am interested in looking more closely at the robe of sanctification that "conservative" wears.

Implicitly if not explicitly, "conservative" claims to be upholding something of traditional, inherited value. It purports to carry a harvest of established standards from the past. It says it represents the accrued wisdom of venerable ancestors and ancient weighty thinkers. Its conventional opposite is "liberal" or "progressive"—words (or political constituencies) that connote a willingness to change, even an eagerness for change, an approach to life or an attitude toward life that may be at times more challenging toward tradition than revering.

Conservative, in the best sense, has a hand-holding relationship with stability. The connection is so obvious that it needs no elaboration. The word "conservation" is illustration enough—as in the "conservation" of nature.

So it is something of a puzzle, and a serious one, how a deadly, overbearing military, tax cuts for the wealthy (not to speak of an economic structure skewed wildly in favor of the wealthy), a nuclear defense apparatus, capital punishment, oil drilling in deep water or the Arctic, scorn for global warming and climate change, refusal to consider universal healthcare, abandonment of small-scale (organic) agriculture and small rural (and neighborhood) schools, indifference to chemical pollution, promotion of nuclear power plants, hatred of socialized mass transit—shall I go on?—how all this and a whole lot more has come to be called or be associated with a "conservative" agenda!

Every single one of the particulars listed above (and I invite you to tuck in your own additional items) is not only *not* conservative, the totality adds up to an assault on *conservare*, on *stabilas*. "Conservatism" is, in fact, the most dangerous political force on Earth, unconditionally committed to the moral presumption of limitless human taking. And that is to say, a great part of our current political disorientation is reflected precisely in the extent to which "conservative" political forces get away with their astonishing assault on Creation, their rapacious exploitation of the working poor, and their ferocious hatred of the common good. By wrapping themselves in the postured dignity of a "conservative" mantle, they practice the precise opposite of what they metaphysically claim to stand for. What we have here is a nearly perfect example of how myth functions.

In this transparent fabrication, we all acquiesce. We all too readily call the party of greed, resentment, and fear "conservative." The media, even the small media of environmentalism and the Left, collude in this semantic shamefulness. Those who would protect Creation are "liberals," "progressives," or "radicals," while those who would log every forest, fish every ocean, mine every mountain, confine every pig, stuff more and more people into "state of the art" prisons, celebrate the "rising tide" of sweatshop exploitation, plan for warfare in space—are "conservative"!

I do not have a handy name for these folks—"deranged greedheads" seems a little too abrasive—but, then, what is there of dignity in their overall proposals and political agenda? One cannot help but affirm "that of God" (as Quakers say) in each of these "conservative" persons, too. But one also hopes—and prays—for their rapid conversion from "conservative" to conservative, from *mythos* to *aletheia*.

But first, it seems, some serious repentance is in order. Repentance, a good dictionary, and a deep, long, refreshing, and stabilizing dip into the Gospels.

Notes

1. Chittister, *Wisdom*, 149.

5

Sacred Cow Crap

It seems a safe and credible thing to say that things are getting worse in the world, crises are building, the old ways of doing things are showing their incapacity to deal competently with reality, and we are floundering in violence and destructiveness. Whether we look at politics, the economy, militarism, healthcare, or the condition of global ecology, what we see are deep and growing political divisions, great disparities in wealth and patterns of consumption, endless talk of weapons of mass destruction (and the need to have, build, or improve them), a raging AIDS epidemic, and a variety of ecological catastrophes from species extinctions to climate change.

So, how adequate, how deep, is our analysis of this mess? I would say not deep enough, and I would also say not deep enough because our field of vision is blocked by sacred cows. Excuse my pushiness as I attempt to shove some of these holy bovines out of our way.

Let me state the obvious. Our view of history, of origins, is culturally conditioned by inherited myth. Our view of the past has been historically dominated by the Judeo-Christian creation story—was and, in spite of Darwinism and other sciences, still is. Our ancient prehistorical past is not something we talk or think much about. Either the ancient past is "pre-civilized," "primitive," or "backward" and therefore not worth consciously internalizing—except, perhaps, in the manner of an entertaining jaunt through a museum or a zoo—or, in the case of six-day creationism, time began five or six thousand years ago and before that there was nothing. (Genesis 1:2 says "a formless void," "darkness over the deep," "God's spirit hovering over the water.")

Sacred Cow Crap

Six-day creationism is one of our very large sacred cows, and the religious Right has an enormous herd of these cows, eager and willing to stampede through any political, educational, or theological discussion. The Right also has televangelist cowboys eager to drive these holy cows right through your living room TV—these holy cows or, perhaps, an unblemished red heifer, as required by Numbers 19, as the animal to be ritually slaughtered for sacrifice in order for the purification rites to be accomplished properly, so that a new Jewish temple can be built in Jerusalem. But not anywhere in Jerusalem: only on the exact site of the previous temples; that is, at the very place where Al-Haram al-Sharif now stands, Islam's third-holiest site, the mosque that marks the place Mohammed supposedly ascended into heaven.[I] (Gershom Gorenberg's *The End of Days: Fundamentalism and the Struggle for the Temple Mount* explores the red heifer of Jewish messianism, the *Left Behind* series of End Times novels by Tim LaHaye and Jerry Jenkins, *The Late Great Planet Earth* by Hal Lindsey, the teachings of Islamic fundamentalists with their anticipations of the End—Jesus, as Muslim prophet, returns to defeat a Jewish Anti-Christ—and any number of other projections of The End of the World that turn eager faces toward Jerusalem and disaster.) If a thousand years are but a day in the eyes of God, and if the longevity of Earth is calibrated to six-day creation, then God's week is in process of coming to a dramatic close.

God made the world in six days. Soon thereafter, Eve and Adam ate of the fruit of the Tree of Good and Evil and thereby committed the Original Sin. God later sent his Only Son to save the world. And, if the world was wonderfully but rather hastily constructed, apocalypse is fully capable of some rapid deconstruction. You had better believe all these things to be absolutely true because, if you don't believe, when you die and are raised for Final Judgment, you are going to be shipped off to Eternal Torment. So

---

I. Retired seminary professor George Wesley Buchanan, writing in the August 2011 issue of *The Washington Report on Middle East Affairs* ("Misunderstandings about Jerusalem's Temple Mount," pages 16 and 64) insists that the site of the Temple Mount has been misidentified for years and that it does not sit at the very place where Al-Haram al-Sharif now stands. Buchanan goes on to say that "Because innocent Evangelical Christians in America, under the guidance of Pat Robertson, Jerry Falwell and John Hagee, have not been informed of these facts, they have thought there was some biblical or religious reason why it was necessary to destroy Islam's third most sacred building in the world, together with the al-Aqsa mosque. It is my hope that, once Christians learn of this mistake, they will stop following Mars and Phineas (Num 25; Ps 106:30–31) and work as zealously for peace, following the teachings of Abraham, the 8th century prophets (Mica 6:8), Jesus, and Paul, as they once worked to promote war in the Middle East. This would make a tremendous difference to Jerusalem—and to the world."

don't say you weren't warned, OK? The End of God's Week is right around the corner.

But the religious Right doesn't have the only pasture for sacred cows. There is another dimension of sacredcowdom in which Right and Left fully overlap. The key word for this vast metaphysical pasturage is—civilization.

Civilization is a huge and sacred bioengineered park sprawling across the political landscape. In the dark depths behind or around civilization's political boundaries lurks a host of vicious wild animals whose names are primitivity, backwardness, paganism, heathenism, underdevelopment, savage, uncivilized, and (we should add the current bogeyman) terrorism. All these came before or are outside of civilization. They are the opposite of sacred cows. They are, to borrow a term from Barbara Ehrenreich's *Blood Rites*, the predator beasts that will kill and eat your sacred cows, if you don't protect those cows with eternal, sacred vigilance. Civilization provides that vigilance. Its surveillance system is second only to God's. God may have avenging angels, but we Americans have predator drones.

Mostly we don't explore the content of these names. We prefer to bandy them about, knock them with political mallets through pre-established lawn hoops in election-year croquet, or ricochet them off church walls in a furious game of hellfire handball. We love to listen to those who can drive these balls straight and true, right through fixed "prophetic" hoops into the safe zone of sacred pastures: home free in suburban homeland security, ready for Rapture.

We love to chitchat about civility, civilized values, having an ambiance of civility, or of civilization itself. These are such warm, secure, cuddly teddy bears. Everything about mama's warm lap and daddy's hardworking reliability is enveloped in these semantic mascots. But what do they really mean? Why, they mean everything warm and nice and neat and fuzzy—if you don't look too closely. So let's look more closely, OK? What do we have to lose but the alleged virginity of our sacred cows and the postured virility of our Viagra bulls?

We are going to have to wade into some elemental definitions and some very basic history. None of this is arcane or outside the easily obtainable public domain. We just don't think about it, don't see it, don't attend to it, because the inflatable sacred cows, so full of sacred hot gas, are blocking our view. So, with sharp etymological pins in hand, let's take a stroll in the pasture of basic history and poke around a little at the inflatable cows.

But before we look at civilization proper, let's look at what we might, with questionable judgment, call the "preconditions" that gave rise to

Sacred Cow Crap

civilization. I say "questionable judgment" because to talk of cultural achievements prior to or outside of civilization as mere "preconditions" is to step into a sacred cow patty and to find sacred cow crap oozing up between our toes—as if civilization constitutes the culmination of all superior aspiration. So let's wash each other's grubby feet and, watching our step, be on our cautious way.

Scholars seem to agree that women gatherers, over time, developed horticulture. Let's say ten thousand years ago, to put a rough date on it. Over a period of generations, increased food abundance led to stable villages, an increased and enlarging human population, a series of new inventions and specializations, and, with the domestication of suitable animals, agriculture proper. One of the scholars I rely on—Lewis Mumford—says these decentralized villages

> ... had perhaps six thousand years to spread via small independent relatively self-contained communities before any attempt was made to unify their activities, increase the tempo of production, or extract the surplus products by coercion or conquest, for the benefit of a ruling class.[1]

This remark appears below Illustration 12: "Neolithic Economy," in Chapter Seven of *The Myth of the Machine*. It implies that the precivilized agrarian village was functioning within a self-contained process of cultural evolution—growing from within—but that civilization, once achieved, acted as a permanent parasite that poisoned the village's evolution.

The real test of our sacred cow crap detector is what we make of the term "civilization." So, with your permission, so to speak, I'll again bring forward Lewis Mumford—same book, end of Chapter Eight. Mumford here compares and contrasts what he refers to as civilization and "civilization." By civilization (without the quotation marks) Mumford means literacy, the visual and musical arts, and the effort to make universal all discoveries, inventions, and creations. By "civilization" (with the quotation marks) Mumford means—and here I quote—the

> ... centralization of political power, the separation of classes, the lifetime division of labor, the mechanization of production, the magnification of military power, the economic exploitation of the weak, and the universal introduction of slavery and forced labor for both industrial and military purposes.[2]

What we have here, of course, is a term, a concept, a word of enormous significance, a word with two utterly contradictory definitions.

The first definition of civilization—art, literacy, the sharing of a universal humanitarianism—can be (and most certainly has been) utilized for their own political benefit by those who are the captains, generals, dukes, princes, kings, pharaohs, emperors, bishops, and popes of the second definition. But let's use one of our sharp semantic pins here and deflate civilization without the quotation marks.

What do we find? What we find within and behind deflated civilization (this is the one promoting literacy, the visual and musical arts, and universal discoveries, inventions, and creations) are cultural impulses and spiritual attainments whose essential humanity and creative wellsprings disappear in the ancient mists—in the so-called "preconditions"—of the Neolithic. That is, these cultural impulses and spiritual attainments are fundamentally and universally *human*: civilization did not invent them. Civilization harnessed them, directed and coerced them, used them for its own amusement and aggrandizement, and then claimed, via its in-house analysts and domesticated proselytizers, to have created them as a kind of spin-doctor "proof" by which to both hide and justify civilization's real functions—which are domination, coercion, and control. As Gil Bailie has shown, such hiding and justifying constitute the core functions of myth. In our time and place, "conservatism" is simply oozing with such hiding and justifying.

So let's do a simple and obvious thing; instead of allowing these critical cultural achievements and attainments to be automatically claimed by or attributed to civilization, let's call them—literacy, art, universal humanitarianism—the manifestations of a profoundly ethical and ancient spiritual *culture* arising from elemental human creativity. Let's attribute these cultural developments to the same process of folk evolution that gave us the precivilized agrarian village, to folk intelligence and common creativity, and not to any "superior" aristocratic coercion.

Well, this leaves "civilization" in the (defined) hands of those who hog power, impose their wills, overpower their neighbors, and exploit the weak. Is this definition unfair? If you consider it unfair, explain, please, how civilization, overtly and proudly so until the industrial revolution blurred the issue, was invariably owned and operated by an explicit hereditary aristocracy and why the "relatively self-contained communities" Mumford sketches (namely the agrarian villages) were perpetually and unremittingly "farmed" by the aristocracy. That is, they were "protected" in a shakedown, mafia-style protection racket sort of way.[II] What I mean

---

II. For an in-context exposition of the word "farm," see "A Landscape Disfigured,"

is, an aristocracy of, at most, ten percent of the total population, lived a life of relative ease and luxury by expropriating the labor and the production of the peasantry, who in turn were compelled to live lives of bare subsistence and grim survival, and then culturally browbeaten for their poverty-stricken "backwardness."

So, let's review a few of the sacred cows: civilization, good; backwardness, bad; aristocracy, good; peasantry, bad. Then here comes the eighteenth century. Here comes the industrial revolution. Here comes French Physiocracy, the origin of contemporary economic theory.[III] Here come massive evictions of the English peasants as the commons are enclosed systematically. Here come huge industrial cities, festering slums, an intensification of European colonialism and the suppression (if not the extermination) of indigenous peoples, a virtual explosion of commercial agriculture, a growing public clamor for democratic governance, and the rise of revolutionary parties of the Left. In this huge social breakdown and cultural disruption, in which civilization dressed in new machinery can easily be seen devouring all forms of noncivilized and subsistence cultures, the agrarian village gets crushed and its crushing—now institutionalized in global economics as normative—is celebrated as a civilized advance, an overcoming of backwardness. Democracy therefore means *civilized* democracy and, according to pundits of Left and Right, can't possibly have any other meaning.

But there's a bit of a problem here. (See, for instance, the July 8, 2002, issue of *The Nation*, with its lead article—"DYNASTIES"—by the left-wing Republican, Kevin Phillips, who, working off data in his book *Wealth and Democracy*, openly warns against the rise of a new hereditary economic aristocracy. The Bushes are the poster family.) The semantic problem is, at its core, an almost totally unexamined one. But it is a semantic problem with fundamentally profound political, economic, cultural, spiritual, and ecological consequences. If civilization in its true historical sense is aristocratic power control, containing elements we can easily identify as both bandit and fascist, then what in the world does it mean to "democratize" civilization? The contradiction is palpable: the alleged democratization of civilization is to make "democratic" a set of governing institutions (i.e., ways of doing things) explicitly designed to perpetuate the suppression of equality, maximize elite control, intimidate and overpower the weak,

---

chapter 10, pages 70 and 71, in my *Green Politics Is Eutopian*.

  III. See "Green Thoughts on Economic Theory," pages 93 through 100, in *Green Politics Is Eutopian*.

worship gods reflective of supremacy (or neuter any that don't), and prevent democratic self-governance from getting serious traction. Go ahead, make a circle of this square while I watch and listen. Explain whether you are democratizing aristocracy or democratizing slavery—or democratizing both simultaneously.

By buying into the aristocratic sacred cow understanding of history—civilization, good; backwardness, bad; aristocracy, good; peasantry, bad—we who think we want simplicity, sharing, an economy within ecological tolerances, universal humanitarianism, and the like, have already lost the argument. By kissing the ass of civilization, by genuflecting in front of the shrine of civility, we have already thrown our lot in with raw power, with a concept of control that scorns and belittles any ecological culture of sufficiency.

Democratic self-governance is only an aristocratic/slavery election game paid for by corporate money if it doesn't impose a ceiling on wealth as well as provide a floor, if it doesn't respectfully attend to ecological consequence, if it doesn't honor and promote democratic, ecological living, if it doesn't recognize, encourage, and protect the ecological resettlement of the countryside, if it doesn't strive for a gentle global humanitarianism. To radically reduce consumption and to radically promote global equity requires a massive sea change of political consciousness, a spiritual movement into a gentle but firm libertarian democratic socialism.[IV]

IV. Naomi Klein is one of our living prophets. In the same article ("Capitalism vs. the Climate," pages 11 through 21) from which I've extracted epigraph paragraphs for part one of this book—"an alternative set of policy proposals," "an alternative worldview," "an interdependence rather than hyper-individualism, reciprocity rather than dominance and cooperation rather than hierarchy"—she also says that "In the rocky future we have already made inevitable, an unshakable belief in the equal rights of all people, and a capacity for deep compassion, will be the only things standing between humanity and barbarism. Climate change, by putting us on a firm deadline, can serve as the catalyst for precisely this profound social and ecological transformation." I would only point out that the "barbarism" Ms. Klein dreads is, in fact, the end product of *civilization's* traumatic institutions in a global mode of breakdown. When enough human beings reach this point of understanding, the "profound social and ecological transformation" will be at hand.

Meanwhile, in the December 2011/January 2012 issue of *The Progressive*, Christopher D. Cook interviews Naomi Klein ("A Progressive Interview with Naomi Klein") on the subject of the Occupy Wall Street rallies. She says (on page 54) "We want to get at the fundamentals of why this system is broken, and why it hasn't just caused the global economy to crash, but is causing all our natural systems to crash as well." Her operative word for "this system" is capitalism. This is true but inadequate and insufficient. Capitalism is a particular, distilled form of civilized economics. It is derivative not original. Its template lies in the marauding banditry that occupied—not Wall

Sacred Cow Crap

We are *not* at the end of a "failed" experiment in democracy. We are at a culminating crisis of globalized civility, a largely pseudodemocracy or infantile democracy saturated with unacknowledged aristocratic presumptions combined with the mutated regimentation of traditional slavery, stampeding its herd of sacred cows all over Earth. Our minds, just like Earth's atmosphere, remain polluted by these toxic gas emissions. Democracy, and the cleaning up of sacred cow crap, has yet to really get underway.

Notes

1. Mumford, *Myth*, 150(l).
2. Mumford, *Myth*, 186.

Street—the agrarian village at the "dawn" of civilization.

Until activists and intellectuals come to grips with this fundamental truth and start saying it out loud and in print, their protest movements will remain shallow and inconclusive. But, as Ms. Klein says, climate change is putting a "firm deadline" on such reluctance. Capitalism has certainly accelerated the entire process of what we might call civilized overshoot. It has done this via industrialization, "development," and "progress." But this means that the entire "lifestyle" of First World populations is now complicit in economic predation and ecological ruination. So it's not just that "capitalism" has to end; our self-identification as civilized human beings (along with an enormous range of what we take for granted as civilized lifestyle entitlements) must be completely reformulated. The stable ethical core of this reformulation process, however, is exactly—interdependence, reciprocity, cooperation—the heart of the matter. It's this ethical groundedness that shapes Naomi Klein's intellectual brilliance into prophetic power.

6

A Left-Wing Odd Duck

KEVIN PHILLIPS SEEMS THE oddest of odd ducks—a left-wing Republican. His *Wealth and Democracy* is filled with charts, graphs, statistics, and facts, the bulk of it detailing, in excruciating particulars, the extent to which the rich in American history have dominated both economics and politics, with the steady deterioration of democracy and democratic vision. He cites six phases, all in the aftermath of war and massive armaments spending, when wealth congealed most rapidly in relatively few hands. These phases (the 1790s, the 1830s and '40s, the "Gilded Age" after the Civil War, the 1920s, the 1960s—and, then, the really big one—the massive wealth transfer of the 1980s and 1990s) have resulted in a concentration of wealth so great that Phillips is openly warning us about the emergence of a new aristocracy. Since wealth concentration correlates closely with power, such concentration naturally bodes ill for the health of democracy.

If I read Phillips accurately, he's both a small-d democrat and a small-c capitalist. That is, he approves of entrepreneurial initiative and private ownership, but he disapproves of enormously concentrated wealth that not only runs the show but ends up being the only show in town.[1]

---

[1]. For an illustration of how this problem of wealth concentration can be resolved in principle, see *The Acquisitive Society* (especially pages 86 and 87) by the eminent economic historian R. H. Tawney or my chapter "A Proper Balance," (pages 157 through 167) in *Nature's Unruly Mob*, a chapter in which I utilize Tawney's prescription for the "abolition of private ownership in those industries, unfortunately to-day the most conspicuous, in which the private owner is an absentee shareholder" combined with the "restoration of the small property-owner in those kinds of industry for which small ownership is adapted"—socialism for the large-scale, in other words, and

A Left Wing Odd Duck

Phillips is probably to the right of Ralph Nader, but he and Nader share an analytical intellectuality, with a capacity for technical detail, that puts the rest of us to shame, or to sleep. My problem with Phillips, however, is similar to my problem with Nader. In neither man do I find an adequate magnitude of cultural depth or spiritual groundedness that tells me I am in the presence not so much of greatness (that's an unfair expectation) as of ample spirituality. I trust both Phillips and Nader with facts—I believe they're both impeccably honest—but their analyses hardly lead me anywhere. Phillips can say that "rollbacks of democracy" have "accompanied the elevation of markets," but I get absolutely no social vision from him of what a fully rolled-out democracy might look like.[1]

Democracy is only another interesting concept, like a weather forecast or last night's bowling score, if there is no vision of what democracy *means*, of what its cultural and spiritual content is or can be. Phillips' vision (if he has one) is tepid.

So what is democracy? Majority rule? Formal elections? Party politics? It seems to me that democracy either means something deeply profound about human fulfillment, about living compassionately and reverently on Earth, or it's just another "lifestyle option" of passing insignificance. If democracy does not mean something indispensably basic about our spiritual groundedness in Creation, about sharing and stewardship (to use again those core terms)—it is going to incrementally collapse into visionless consumer stupidity—it already is—until brazen aristocracy once again shamelessly asserts its "divine right" and assumes full global control.

One has to acknowledge (Robert Caro's biography of Lyndon Johnson comes to mind) how the studied pretense toward democracy on the part of many politicians has served to corrupt, deeply and powerfully poison, the very meaning of democracy: the naked ambition for public prestige and high office, for wealth and power, the shameless pandering and brazen lying, the outright contempt for democratic process hidden behind a manipulative patriotism and an idolatrous use of religion—all of which makes "politics" a filthy word.

Democracy means, at its heart, that we are and recognize ourselves as creatures of mystery, beings whose very existence as self-conscious persons is totally inexplicable, whose task is the global unfolding of sharing and stewardship, the reciprocal deepening of our souls through compassion and reverence, tenderness and forgiveness. But if we can't—or won't—talk

---

private ownership of the small-scale, with the latter so numerous and vigorous that the former is kept clean, efficient, and uncorrupted.

about this core, this heart of the matter, we are left with process, with procedure, with questionable precedent, with extremely bright guys like Kevin Phillips and Ralph Nader—honest, trustworthy, earnest—but with an inadequate spirituality.

The crisis we are in is only secondarily a political crisis. Our primary crisis, a crisis that bears an enormous burden of accumulated distortion and sin, is spiritual in nature. Democracy will fail if we refuse the task of this inner growth. I like Phillips. I like Nader. I'm even intimidated by their command of detail. But I also want to grab them by their conventional businessman ties and demand that they get out of their abstracted heads and begin speaking from their souls.

Democracy is not a concept that needs "cyclical" renewal. Democracy, in its true and fullest sense, is the spiritual exploration of servanthood and stewardship, an exploration we have hardly begun to engage, an exploration that absolutely depends on the softening of our hard hearts. At the molten core of democracy lies the same energy that animates the kingdom of God. With the invocation of this "kingdom"—it can be totally nonreligious but it simply must be spiritual—democracy can awaken to its creaturely potential and become servant gardener to the Earth.

Notes

1. Phillips, *Wealth*, 417.

7

What Is "Religionless Christianity"?

"Village-mindedness" is one of the dominant concepts associated with the life and career of Mohandas Gandhi. In the case of Dietrich Bonhoeffer, the term for which he is most famous has come to us in English as "religionless Christianity."

On April 9, 1945, Dietrich Bonhoeffer was hanged by the Nazi regime. He was thirty-nine years old, a Lutheran pastor and theologian. His "crime," in the great Christian nation of Germany, was to oppose Hitler so completely that he became a participant in an assassination plot. For this, he was arrested, imprisoned, and hung.

Bonhoeffer is remembered and celebrated for his boldness and courage—one of the few who refused to go with the flow of pathological German nationalism, a nationalism seemingly concocted out of irretrievably damaged empire pride blended to perfection with cunningly stimulated resentments—but also for his realization that Christianity as a religion was largely dead. Two world wars in his short lifetime—wars instigated and fought primarily by European "Christian" nations—were powerful evidence for such a mortal view. Nationalism and supremacy wrapped in resentment were accrued diseases that profoundly corrupted love and undermined stable bonding with Spirit. The nation-state, operating under Machiavellian principles of *Realpolitik*—craftiness, duplicity, the devious hypocrisies of power—had become openly genocidal and even ecocidal in its accumulated technologies and standardized organizational structures, now capable of destroying the world. Or, if not "the world" in a total sense, then capable of exterminating a huge variety of life forms on Earth, most

The Kingdom of God Is Green

certainly including ourselves. The civilization that achieved this magnitude of suicidal and ecocidal power was and is explicitly "Christian."

Yet Bonhoeffer was not given to despair. He saw the corruption of Christianity as part of the modern world's "coming of age," a world trying to crawl out of the mummy wraps of its religious swaddling clothes. The world, he said, has outgrown the pieties and platitudes of sanctimonious religion. Instead of being scolded or browbeaten for irresponsible delinquency, humanity should be encouraged to get off its hands and knees and learn to walk.

In the fragments of letters and essays surviving Bonhoeffer's final months in prison, one term, one concept, really stands out—"religionless Christianity." For nearly seventy years, various intellectuals and countless theologians have grappled with that construct, trying to see what Bonhoeffer saw (or at least glimpsed) in the extremity of his apocalyptic circumstance, and trying to elaborate or build upon that vision—but, of course, without the immediacy, sheer psychic stress, spiritual upheaval, and remorseless political pressure cooker that was Bonhoeffer's world.

I am by no means familiar with the totality or even the bulk of these grappling efforts.[1] But what I have seen of those efforts suggests a rather frustrating fascination with the concept of "religionless Christianity," a fascination held in check by theological convention and largely confounded by religious swaddling clothes, rather like a resuscitated mummy trying clumsily to wiggle through its moldy rags or maybe trying to rewrap itself. That is, the impulse seems to be to pull this intriguing idea into prevailing institutional understandings, to "integrate" it into pre-existing creedal convictions, to somehow taffy-pull "religionless" comfortably back into religion, to helpfully tug Bonhoeffer back from the brink, all the while patting him on the back for being such a brave and even prophetic soul. But if my hunch is correct, such efforts—perhaps while popular and even, in a pious sense, "cutting edge"—are doomed to failure.

Why such trouble? Why so much churning in sand? If we have difficulty (as most of us do) getting our minds even remotely around "religionless" Christianity, let's back up a bit to realize that it was Christian flabbiness and flaccidity on the part of the German churches that enabled Hitler to exercise such wanton power and awesome destructiveness. If Germany was overwhelmingly a *Christian* nation, composed primarily of

---

1. Recently I came across a book by William Kuhns, published in 1967, called *In Pursuit of Dietrich Bonhoeffer*. Kuhns, a Catholic scholar and layman, was clearly writing sympathetically in regard to Bonhoeffer, while also in the uplift of Vatican II. It is a serious and sensitive book, very much worth reading.

What Is "Religionless Christianity"?

Catholics and Lutherans in roughly equal proportion, how in the world could Hitler happen? How could Nazi ideology take possession of such a Christian country? How could national religious piety produce (or allow) such a monstrosity?

There are many layers here. Shall we say that virtually all modern Christians have become flaccid and flabby? (Soren Kierkegaard was shouting this news in Denmark a century before Bonhoeffer.) But are contemporary Christians any more flaccid and flabby than those, say, at the time of the (more or less) genocidal Crusades, any less given to self-satisfaction, mass hatred, and collective hysteria? Could we say that when Christianity locked itself into being the State Religion of the Roman Empire it took an oath of perpetual flabbiness? (If the Roman Empire was the primary murderer of Jesus, what do we call it when his "Bride"—the church—enters into permanent concubinage with his killer?)

All of this is to say, if we have trouble getting our minds around "religionless" Christianity, perhaps we need to balance (or compound) our mental troubles by embracing a complementary concept. Let's call it—even though it hits us, as "religionless Christianity" hits us, as oxymoronic—"noncivilized governance." Perhaps it's simply not possible to get traction on "religionless Christianity" without recognizing the bipedal necessity of having another conceptual leg to stand on. If it's not possible to understand what Christianity explicitly became (or had become) at the time of the Constantinian Arrangement without simultaneously understanding the nature of the beast to which it became attached, what makes us think we can get a handle on *religionless* Christianity without a comparable grasp of *noncivilized* governance?

Let's remember that theocracy is the formal merging of church and state, while the idea or doctrine of "two kingdoms" is theocracy split in half. (Or, if theocracy is, as *Webster's* says, the "rule of a state by God or a god," then the doctrine of two kingdoms may reflect the hemispheres of God's brain, one side of which is given to love for the chosen few and the other to wrath for the unchosen many.) A major conceptual transformation in one hemisphere therefore implies or requires a major conceptual transformation in the other. To talk of religionless Christianity without acknowledging or intellectually delving into its conceptual other half is bound to result in confusion and befuddlement. It's an incomplete picture, an unfinished diagram. "Religionless Christianity" requires "noncivilized governance." They are, in our Augustinian world, a dialectical pair.

The Kingdom of God Is Green

Religionless Christianity implies, at least, a spiritual delinking from creedal theism. But noncivilized governance also implies a cultural delinking from the Class-and-War basis of imposed globalization, the sort of civilization Lewis Mumford wanted kept in quotation marks. If the world's "come of age," then the spiritual implication of globalization is the readiness of Spirit, in part via the infectious contagion of Gospel yeast, to risk an awesome transformation of human consciousness and behavior in the modes of ethical religionlessness and noncivility. This seems to mean a far more mature level of spiritual comprehension, fully inclusive of human self-governance and ecological economics. Perhaps Spirit has gotten sick and tired of our willful and irresponsible spiritual immaturity, our utopian hubris, and thinks it's time for human beings to grow up and act responsibly. Put up or shut up. Grow up or go extinct.

Or, to put it differently, we have to recognize that theocracy expels the kingdom of God to a far distant (and largely otherworldly) eschatology. It overrules, blocks, and thwarts "Your kingdom come on Earth." The doctrine of the two kingdoms is a fundamental violation of the Lord's Prayer.[II] So while the discovery of "religionless Christianity" requires a comparable discovery of "noncivilized governance," the implosion of theocracy serves to restore temporal energy to the kingdom of God, that woefully neglected "concept" so prominent in the Gospels. The empowerment of the kingdom of God also implies a fusion of humble spirituality and ecological self-governance; but it does so from within a "kingdom" or folk or ethical perspective rather than from an otherworldly theocratic or civilizational or doctrinal perspective: evolutionary not dictatorial. And that's why the kingdom of God means libertarian democratic ecological socialism—or, more simply, why the village-mindedness of God is Green.

II. In Roland Bainton's 1950 book on Martin Luther—*Here I Stand*—Erasmus is described, on page 253, as asking Luther "whether the ethical precepts of the Gospels have any point if they cannot be fulfilled. Luther countered with characteristic controversial recklessness that man is like a donkey ridden now by God and now by the Devil, a statement which certainly seems to imply that man has no freedom whatever to decide for good or ill." And, while Bainton documents Luther's insistence on the Bible as the touchstone of religious certainty (*sola scriptura*), there is in *Here I Stand* absolutely nothing on the meaning or significance of the kingdom of God proclamation, a fact that silently underscores Luther's captivity within Augustinian metaphysics.

8

The Aristocracy of Consumption

THE POLITICAL REVOLUTIONS OF recent centuries—the American, French, Russian, Chinese, and Cuban, to name a few—have had at least one common element. That element was an effort to contain, curtail, or even overthrow a pattern and a practice of aristocratic governance and to establish a pattern and a practice of governance based on popular consent or, at least, governance in behalf of "common" people. Beginning in the late eighteenth century, "commoners" began to assert (and to some extent achieve) a broader right to self-government.

The difficulty of first envisioning and then establishing a form (or forms) of "common" self-governance, freed not only from the actual personages of aristocratic presumption but also from the metaphysical preconceptions of civilizational presumption, has rarely been appreciated or understood. "Democratizing" the inherited governing structure (or structures) of civilization seemed self-evidently sufficient. Add onto this difficulty (Robert Caro's multivolume biography of Lyndon Johnson provides a painful illustration readily at hand) the very wide-spread phenomenon of persons seeking public office not on the basis of committed servanthood to democratic principles but out of lust for power, popular acclaim, notoriety, and cheap glory. (As Caro says of Johnson in *Means of Ascent*, "Not only did he not want to be regarded as an idealist, or as a fighter for causes, he wanted to be regarded as a man who scorned ideals and causes as impractical dreams, as a man practical, pragmatic, tough, cynical. He wanted the world to see him not as merely smart but as shrewd, wily, sly.")[1]

That each of these political revolutions named above was only partly successful in carrying out its stated objective, given the historically embedded religious, cultural, and psychological massiveness of utopian presumption and aristocratic prerogative, the sheer hegemony of civilizational preconceptions, and the enormous difficulty of practically working out what "democratic self-governance" actually means in daily life even as undemocratic vested interests invariably attempt to bend policy decisions in an undemocratic direction (including the flagrant use of patriotism to achieve undemocratic ends)—all this really shouldn't surprise us. Compound the practical difficulty not only with cynical and sly manipulations of democratic aspiration but with vacuous assertions like this one from Kevin Phillips' *Wealth and Democracy*—"Whereas liberal eras often fail through utopias of social justice, brotherhood, and peace, the repetitious abuses by conservatism in the United States in turn involve worship of markets (the utopianism of the Right), elevation of self-interest rather than community, and belief in Darwinian precepts such as survival of the fittest"[2]—and it's easy to see why the implementation and deepening of democratic self-governance is so difficult. (I call Phillips' assertion vacuous because of its slurring of "social justice, brotherhood, and peace" as utopian and liberal, as traits supposedly leading to failure. With friends like this, democracy hardly needs enemies.)[I]

Yet we often are surprised to hear, for instance, criticisms of democratic procedure (or lack of it) in American society, as if we were astonished to discover someone saying that "our" revolution was less than totally complete, that it only just opened and most certainly did not close the exploration of democratic or popular or, simply, common self-governance as a new development in the history of civilization. Our elementary school understanding of democracy is similar to our Sunday school understanding of Christianity. We had a great revolution toward the end of the eighteenth century, complete with heroes and heroic efforts, and, with the possible addition of the Emancipation Proclamation and women's suffrage, there's basically nothing more to do except protect The Greatest Society That's Ever Existed In The History Of The World from evil terrorists and their insane jealousies. In the case of Christianity, Jesus did it all for us two thousand years ago; all we need do is believe he did it, perhaps aided

---

I. On page 371 of *Wealth and Democracy*, Phillips expands on his critique of utopias by saying, "The utopia of American liberal or progressive politics has been the perfectibility or the achievability of justice and equality. The equally unachievable utopia of economic conservatism has been laissez-faire or the perfectibility and enthronement of the market." This critique may well be true in regard to utopianism; but eutopianism is far less interested in any sort of metaphysical perfectibility—whether of human nature or the capitalist market—than it is in life lived wholesomely within ecological limits and ethical understandings.

The Aristocracy of Consumption

by the Reformation (if Protestant) or Counter Reformation (if Catholic). With American democracy, it was the Founding Fathers who brought us political completion if not perfection. In both politics and religion, in other words, it's a done deal. All we have to do is "believe" it's so and ride along on the achieved energy of their astonishing accomplishments.

This is an interesting position, but hardly a great or a profound one—except great in its myopic "innocence" and profound for the deadly mischief attendant on the enactment of such "innocence" in the larger world.

The problem with the overthrowing of civilizational governance lies in the great and subtle difficulties in becoming aware of all the ways, means, devices, and subtleties by which aristocratic bias has permeated cultural consciousness, on the one hand, and, on the other, finding our way into a truly adequate alternative to the aristocratic worldview. We believe we live simultaneously within a Christian civilization and a democratic civilization; and that means simply working our way through the illusions created and sustained by these semantic conventions is the first very difficult step.

Since Constantine's time, more or less, the church's job in relation to the state was to bless the catapult and kiss the sword. It was to be the "spiritual content" in an essentially aristocratic drive to achieve "Christian civilization"—with Christian in a suitably adjectival role.

So what happens, several centuries later, when popular revolutions make at least a tiny inroad into the hegemony of aristocratic governance? What does the church have to say about such things? How does the church understand the nature and function of democratic self-governance? What does the church mean (if it means anything at all) by the "kingdom of God"? Does the church recognize any connection whatsoever between the struggle for democracy and the spiritual unfolding of the kingdom of God?

Perhaps it's best to say, at least provisionally, that the church's understanding of the kingdom of God is about as weak and underdeveloped as the citizens' general grasp of democratic self-regulation. Both these realities—the kingdom of God and democratic self-governance—require a depth and magnitude of spiritual maturation that our childish adherence to political cliché and religious dogma invariably forestalls and obstructs. We are still governed, in the very language, images, and ideas that shape our thought and guide our behavior, by inherited patterns of deference to aristocratic prerogative and utopian presumption. We have not so much democratized civilization as we have democratized slavery. Does that seem outlandish and outrageous? Allow me to dig that hole deeper.

Perhaps a basic economic distinction between civilized and noncivilized social organization has to do with "surplus" versus sufficiency. As a generality we can say that all precivilized or noncivilized people lived lives of sufficiency—producing essentially no more than they needed. This implies an elastic cultural embeddedness of self-regulated leisure and a general absence of externally imposed coercion—for why, lacking coercion, would any person or any society deliberately produce more than was necessary for the maintenance of life, unless there was some sort of pathological obsession?

The historic link between sufficiency and "surplus" is agriculture. That is, as gatherers in the late Stone Age got more adept at producing grains by encouraging and enabling greater plant concentration, they opened the door to an absolutely new era of abundance: horticulture. First the domestication of plants, then the combined domestication of plants *and* animals: agriculture. What's peculiar to our common understanding of the agrarian experience is that, on the one hand, we are ready to shout "Abundance!" with the discovery and early practice of both horticulture and agriculture; yet, on the other hand, we have a sour conviction that the peasant experience was always and everywhere mired in "scarcity." How do we reconcile this enthusiasm for agrarian "abundance" with anxiety toward peasant "scarcity"?

As a rule we don't reconcile this contradiction between abundance and scarcity because a true reconciliation entails recognizing the structural massiveness of the civilizational crime of armed theft, a recognition with subsequent institutional confession and historical repentance. We are inclined to confess only insignificant sins and evade the ones that really matter. We are, as persons within a civilized *system* governed by self-interested elite presumption, both discouraged and prevented from looking too closely into this abundance/scarcity contradiction. Our history texts celebrate the spread and victory of civilization; insofar as the economic foundation of civilization is explained at all, it is largely explained away. Never is it acceptable in conventional society to assert that civilization came into being with slavery and extortion and that, in disguised and to some extent ameliorated form, it continues to exist on that identical basis. Capitalism is "democratic" civility. That is, it dangles the prospect of affluence and riches before everyone, regardless of class or origin, and celebrates the accumulation of commodities and wealth as both a civil and religious virtue—in total contradiction, needless to say, of all Gospel teaching.

The key transitions to the modern system occurred from the late eighteenth to the early twentieth centuries: a combination of economic

The Aristocracy of Consumption

and political revolutions. These are, in brief, the massive European eviction of the peasantry from the commons, the rise of the capitalist factory system, the political overthrow of explicit aristocratic regimes, the deepening institutionalization of global European colonialism, the relentless commercialization of agriculture, and the struggle (always against entrenched interests) for a deeper understanding of and broader participation in democratic governance on the part of the lower classes.

Soviet communism, for a major example, was a huge effort to impose industrial civilization on a demographically "backward" peasantry in the name of the proletariat: but really a utopian preconception implemented by a ruthless dictatorial elite, a recapitulation, as Lewis Mumford has shown, of the very earliest civilizations. That is, the infection of civilized preconceptions, deriving from the affluence, elegance, and philosophic formulations of classical aristocratic civility, provided the intellectual framework in which even "democratic" aspirations were to be achieved. Revolutionaries may have earnestly desired to build a "modern world," but their blueprints and tool kits were explicitly civilizational, always robbing the countryside in order to enlarge the city and magnify the nation-state.

What I'm trying to articulate is both very simple and very difficult. True democracy, at least in the language of Gospel spirituality, must pass through the spiritual filter of Jesus' "kingdom of God" in order to be truly whole and wholesome. But the origins of civilization lie in the aristocratic expropriation of agrarian abundance, an *institutionalized* expropriation, and those origins and expropriations are with us yet, only scientifically intensified. The historic maintenance of civilization up until the industrial revolution depended upon some form of slavery or explicit human economic bondage. (The aristocracy always "governed" the peasantry both politically and economically.) But with the political and technological revolutions of the last two centuries plus, we somehow imagine (have been led to believe) that these "revolutions" magically democratized the complex body of inherited undemocratic institutions and aristocratic behaviors we call civilization, and, furthermore, that this "democratization" is, for all practical purposes, indistinguishable from consumer affluence. If we've got lots of stuff, we must be "free."

There is in fact for many people, especially First World people, hugely more freedom of choice, more life options, than was ever true of past generations. So it's not as if "greater freedom" is totally delusional.[II] Material

---

II. See Chapter 4, "Absolutely Scandalized," in my *Polemics and Provocations* for a look at the underlying tensions and unresolved conflicts between civilization and folk culture.

surfeit without spiritual, political, or cultural groundedness results in the anomie and alienation so pervasive in modern life. The erosion of folk culture is now fully realized. We have all been processed through civilized institutions. Commercial emptiness and mass loneliness are the norms of daily life. In this state of spiritual despair, democracy becomes anemic. Citizens are as if drugged and in a stupor, unable to make coherent sense of political or religious rhetoric. How did this happen?

If we look structurally at, let's say, agriculture or rural culture as a whole, we find the great human bulk of agriculture transmogrified into technological agribusiness and the vast cultural complex of rural culture shredded into virtual nonexistence. The price of "democratic" civilization has been the utter destruction and not the democratic transformation or resurrection of the peasant village. Instead of restoring economic coherence to the agrarian village, industrial "democracy" destroyed the village. If the consolidation of civilization suppressed cultural evolution in a folk sense, then one could logically anticipate that democratic revolution would lift that suppression by restoring the coherence and freedom of the agrarian village. Needless to say, industrial "democracy" did just the opposite. Furthermore, we implicitly if not explicitly rationalize this negative orientation toward agriculture by a bias against "backwardness." That is, the peasant village (and, in the end, small-scale rural culture as a whole) was identified as the repository of "backward" cultural elements restraining humanity from civilized advancement. As soon as it was technologically possible to do without the peasantry, civilization jumped at the opportunity. Industrialization landed on rural culture with both iron feet.

What we carelessly overlook is that the aristocracy supplied the long-term chokehold on the peasant village. The aristocracy not only "governed" the peasantry, it also taxed and expropriated all possible "surplus" and kept the peasant village in a state of economic scarcity and cultural constraint. Yet when the opportunity arose, in principle, to act "democratically," it was the peasantry who got hammered, and aristocratic prerogative was transferred to the get-rich mass aspirations of a capitalist system, both as owner and consumer. The culture of agrarian life died, food became dirt cheap (especially at the raw commodity level), and the entire "civilized world" rose to an induced aristocracy of consumption. The "civilized world" as a whole now treats the Third or Fourth World in a manner directly analogous to how the aristocracy treated the peasant village historically: sucking out its "surplus" by means of violence, expropriation, and debt, while hammering it simultaneously for "backwardness" and "excess" social spending. We have become "democratic" aristocrats, and we are ethically numb because of the imperialist surfeit with which we are complicit and to which we feel entitled.[III]

In other words, our democratic institutions and political rights, critically important as they may be as opening victories in an ongoing process of vastly deeper and deepened democratic transformation, are not only partial and limited achievements, they are also deeply stained (especially the institutions) with aristocratic infections, constantly corrupted by capitalist consumption and intimidated by patriotic flag-waving.

In our "democratic" rush to overcome "backwardness," we crushed the victim (the peasant and the indigenous) while aping the values of the victimizer (the civilized aristocrat). Maybe that's the phrase I'm looking for: "democratic" civilization is aristocracy's ape. If so, then the global crisis we are in can only be resolved by a profoundly humanitarian and spiritual transformation that restores the agrarian village to both its richest ecological coherence and its fullest democratic potential. The Good Shepherd is the sower of these seeds.

Notes

1. Caro, *Means*, 7–8.
2. Phillips, *Wealth*, xxi.

III. With some hesitation and even trepidation, I say this presumption also exists among the public sector workers whose unions, pensions, and wages were hammered in Wisconsin by the right-wing political machinations of Governor Scott Walker and the Republican-controlled legislature in 2011. That is, I believe most of the First World is complicit in economic "overshoot," certainly including the great bulk of the comfortable middle class. I'm not saying, however, that public sector workers have no right to fight back against the right-wing onslaught—which, at its core, certainly seems to be an effort to concentrate yet more wealth and power in the hands of the few—but workers also have the moral obligation to recognize overshoot (both its economic and ecological consequences) and to help devise policies that bring human life within the ethical framework of servanthood and stewardship. When Christian Parenti, in his *Tropic of Chaos: Climate Change and the New Geography of Violence*, can use phases like the "politics of the armed lifeboat" and "climate fascism," as he does on page 11 of his alarming book, we have to realize he's talking about the United Sates above all, and that unmistakably implicates the affluent lifestyle to which we've all become attached and to which we feel entitled. That the Right cynically uses deindustrialized working-class resentment to "justify" its attack on public sector unions does not mean that middle-class affluence is exempt from critical analysis in regard to overshoot. But, if anything, the Right's repulsive tactics have made middle-class self-examination even harder to achieve because of heightened defensiveness.

9

The Kingdom of God Is Green

HERE IN WISCONSIN, OUR North-Central Worship Group, just over a dozen people, is under the care of Madison Meeting, though "under the care of" is, at best, ambiguous, for we are roughly 150 miles north of Madison, and the radius of our scattered Worship Group is, easily, fifty miles. We don't see much of Madison and Madison doesn't see much of us. Plus we local Quakers don't see enough of each other.

Few of us—very few—are "birthright" Friends. Most of us have migrated into Quaker practice from the Christian mainstream. It can even be said that we have found in Friends Meeting a refuge from an unpalatable Christian convention, a convention largely made of polished mythology juxtaposed with ethical evasiveness.

And so we struggle, each and all in our own way, to grasp what we are to do, what we are to expect, how to achieve a patient and humble internal quietness in our hour of silent worship as we wait for the stirring of Spirit. Opening to one another our respective muddles in regard to the meaning of "meditation," in Meeting for Business, we have decided to try an experiment. At the beginning of each Meeting for Worship, we will, each in turn, bring something—music, a reading, a talk—that's not to last longer than fifteen minutes and is to facilitate a state of inner quietness or provide a focus for contemplation in the ensuing silence. I obviously have chosen to give this talk; and I wish to bring what I believe are powerfully related dynamics that impinge mightily on our capacity to live deeply meditative lives. I will, of necessity, be brief.

Marcus Borg, in a book entitled *The God We Never Knew*, describes in some detail several images of God, and he explains how and why these

The Kingdom of God Is Green

differing images matter. The dominant image of God in Western Christianity, Borg says, is the "monarchical model," an image of sheer transcendence, God as king, judge and law-giver, in relation to whom we are to be obedient peasants.[1] This God is male, distant, powerful, stern, and to be feared.

This image of God represents a projection into divinity of the dynamics within strictly hierarchical empires and kingdoms. This God is "King of kings." Although this imagery obviously has ancient roots, inclusive of that historical and mythological context we call the Old Testament, its congealing event for us was the merging of secular kingdom and Christian faith in the Roman Empire reign of Constantine.

Almost contemporaneously with Constantine's forcible "unifying" of Christian doctrine, there emerged, largely as mute protest, the great monastic movement of Western Christianity, beginning with the Benedictines. Rather than rote exterior obedience to "divine" authority, monks and nuns institutionalized the inward life, asserting that to experience and know God it was (and is) necessary to find good quiet soil to grow in, in an environment not choked with distracting and debilitating weeds.

Augustine, meanwhile, in *The City of God* (see especially Book V), was instructing the Western world that kingdoms and empires are given and sustained by God. This assertion pervades and saturates Christian doctrine to this day, and it has formally evolved as the doctrine of two kingdoms. The Church is one arm of God's kingdom. The State is the other. With the first arm, God offers eternal salvation and, with the second, He keeps order on Earth among unruly humans.

I propose to you that this teaching defiles, corrupts, and undermines that "concept" propounded in the Gospels by Jesus—namely, the "kingdom of God" (Luke) and the "kingdom of heaven" (Matthew). It is my conviction that scholars like Marcus Borg (*Meeting Jesus Again for the First Time*) and John Dominic Crossan (*The Historical Jesus: The Life of a Mediterranean Jewish Peasant*) are showing us how socially revolutionary Jesus' teaching and practice were.[1] What Jesus was proposing was a "yeasting" of servanthood and stewardship in human consciousness or human spirit by which *the entire process of human life* would be overtaken and transformed into the "kingdom of God."

I. Perhaps even more gritty and down to Earth—certainly more existentially scary—is Ched Myers' methodical unfolding of Jesus' wandering trajectory toward his eventual murder in *Binding the Strong Man*.

The Kingdom of God Is Green

It is, in one respect, incredible that everybody doesn't know, or grasp, that civilization killed Jesus. We have been told the story of his crucifixion so many times that it has become simply that—a story—a sad and moving one, perhaps, but not at all instructive for our present life and political circumstance. Our primary conception of civilization, thanks in part to the doctrine of two kingdoms, is a rosy one. If pushed ever so slightly, we will assert that civilization is inherently good, that it has raised us from subhuman into human, that it actually spiritualizes us in a transition from "primitive" to "civilized," and that we can in no way do without it. We can't imagine anyone becoming "saved" without first (or at least simultaneously) becoming "civilized."

I propose the opposite. I suggest that just as civilization killed Jesus, his murder by the collusion of religious and state authorities reveals the repressive stance civilization takes toward the kingdom of God. That is, civilization is about the maintenance and expansion of power, wealth, fame, and advantage, all of it aligned beneath the "will" of the monarchical image of God, in varying formations of "divine" hierarchy. This functioning paradigm scorns, and it will crush, attempts to formulate social relations based on powerlessness, simple sufficiency, modesty, and sharing. This crushing spans the spectrum from legal murder to ecclesiastical triumphalism.

Why is it important to think critically of civilization and the kingdom of God? Why do they matter? In regard to the latter, the answer is easy and simple: it's important to know, as fully as we are able, what Jesus meant by the kingdom of God because it's the term he used to talk about something very important, if we are to believe the Gospels. In regard to the former, the answer is harder and more complex. On the one hand, we float in a pervasive mythology that associates civilization with nearly every human advance and technical refinement for the last five thousand years. In this assessment, civilization is the name, the term, that exemplifies these refinements and advancements both as motive force and as achieved ideal. On the other hand, especially as seen, for instance, in the careful analysis of world historian Lewis Mumford, civilization arose as authoritarian kingship by forcibly expropriating the abundance of Neolithic agrarian villages, by militarizing those nascent kingdoms for purposes of internal control and external conquest, by institutionalizing slavery while rigidly enforcing a vast social cleavage between the aristocracy and the peasantry, and by authorizing formal religious doctrines (with an appropriate priestly class) that demanded submission to the king's divine will.

Although the last two centuries have seen democratic upsurges take shape around the world—both reform movements and revolutions—essentially all of these efforts have operated within the parameters of inherited civilized values, producing in the end variations of the same debilitating authoritarianism and wealth concentrations they initially vowed (or at least implicitly promised) to overcome. Insofar as these reform and revolutionary movements were fueled by an intention to undo aristocratic control of human life, they have failed to achieve their goal, for the disparity in wealth is as great now, proportionally, as it has ever been in human history. While some of these attempts at reform and revolution were certainly influenced by the radical values of the Gospels, none, except for social movements led by people like Mohandas Gandhi and Martin Luther King, had the spiritual focus and strength of will to both intellectually grasp and politically strive for the kingdom of God as political "ideal." The most radical political movements have been those associated with explicit spiritual content and guiding spiritual energy.

If we look closely at the doctrine of the two kingdoms—the Church as agency of otherworldly salvation, the State as agency of earthly order—it's fairly easy to conclude, if we've read the synoptic Gospels with care, that the two-kingdom doctrine has virtually nothing to do with Jesus' teaching and practice of the kingdom of God, which is overwhelming about loving Spirit with all our might and loving our neighbor, even the neighbor who is our enemy, as we love ourselves. *Realpolitik* is not a core value in the Sermon on the Mount. Machiavelli is not a Franciscan saint.

To follow with single-minded passion these two teachings—to love Spirit with all your heart, and to love your neighbor as you love yourself—leads to either civilization's end or to the transformation of civilization's institutions from principalities of predatory power into humble agencies for the exercise of servanthood and the exploration of stewardship. Either way—the "end" of civilization or its thorough transformation into something simultaneously humane and ecologically reverent—would result in a radical shrinkage of the monarchical image of God.

I will attempt to bring this back to the matter of meditation or, more significantly, to the matter of living more meditative lives.

The Book of Acts suggests there was a fairly strong movement, after the legal murder of Jesus, for the new "Christians" to live in community, sharing wealth, homes, and possessions. With the later monastic movement this impulse is explicit and historically continuous—except that it was, and is, explicitly for celibate individuals and not at all for married people or families. That is, the shadow side of the monastic movement resulted in a terrible thing: spiritual community, in its fullest sense, excluded married people and families. The kingdom of God, it seems, was only for professional celibates. Sex was quarantined as if it were a pest or a disease.[II]

The Protestant Reformation, by and large, instead of working to correct this glaring spiritual deficiency, compounded it by abolishing monastic life altogether. No longer, in the Protestant world, was it possible to retreat from "the world" into a deliberate spiritual life of community and cooperation where the *ethical* elements of human economy could be carefully examined and shaped. Only the radical Anabaptists made a decisive effort, ongoing to this day, to form explicit spiritual communities composed of entire families with a determined orientation toward sharing and modest sufficiency. For the most part, the contemplative, meditative, and mystical aspects of spirituality were utterly neglected by mainstream Protestantism in favor of Bible-based salvation exhortation. Aside from exhortations to "live a moral life," there were virtually no teachings pointing toward or enabling one to move toward the kingdom of God, no barriers in the way of maximizing personal or private wealth. The function of religion was to save individual souls for the heaven to come. This life was under the jurisdiction of the State. To contemplate the possibility of the kingdom of God in this life—except, perhaps, for chaste monastics—was to dabble with heresy.

There is no evidence, however, that Jesus suggested that the kingdom of God was available only to cloistered celibates. There is no evidence that Jesus concurred that this life necessarily falls under the jurisdiction of the State or that this jurisdiction is the will of God—or that sexual appetite is of its nature evil. There is every reason to believe that Jesus' vision of a spiritual social order defied the rigid, extractive, controlling legalisms of an oppressive system of repressive civility, and, instead, proposed vital communal living with radical sharing and thoughtful simplicity as that sort of life most fully characterizing, or at least compatible with, the kingdom of God on Earth.

II. In his chapter on Augustine ("An Unloving View of Sex"), Peter De Rosa, on page 319 of *Vicars of Christ*, says "The holy fountain of life, [Augustine] said, was always dirtied by lust (*libido*) even in the tidy garden of marriage. His conviction that sexual appetite is of its nature evil became the great tradition of the church. Because of the Fall, man has been attacked at his most vulnerable point: sex." It is therefore not surprising that, in Luther's day, according to Roland Bainton (*Here I Stand*, pages 33 and 45), monasticism "was the way par excellence to heaven" or that the cloister was the form of "higher righteousness." Nor is it incidental that Luther was an Augustinian monk.

The Kingdom of God Is Green

It may well be that some people have achieved, and others will achieve, a mystical union with Spirit via a discipline of meditation and prayer, of concentrated inner resolution. Sidestepping here the question of how we imagine God, of what image we project onto God, I will only say we all, it seems, at times, desire this mystical union. But I think it's true that only rarely do we desire this union with unrelenting passion. Our generalized "intention" may be good and sincere, or so we tell ourselves, but we are weak and immature plants often choosing to be choked with very interesting and entertaining weeds. And that brings us to the question of how we can live lives in which our spiritual soil is richer, where other spiritual plants, healthy and strong, are close at hand, where we have disciplined ourselves (and each other) from an inordinate affection for choking weeds.

The answer, I believe, lies in taking the kingdom of God far more seriously than we do, recognizing our bondage to the perniciously repressive (and paradoxically permissive) strictures of the two-kingdoms doctrine, learning to disconnect from the prevailing religion of civilization-worship, while earnestly and prayerfully moving toward living in simple, ecological community.

We cannot, except in rare moments of insight or grace, live meditative lives if our allegiance, in the practical conduct of our everyday lives, is shaped and determined by our belief in (and dependence on) the predatory State, with its formal economic ideology of progress, growth, and the essentially limitless expansion of personal consumption dictating our actual behavior. This is, I believe, what Jesus meant when he said it's either Mammon or God, but not both.

We are dabblers in meditation because we are dabblers in the kingdom of God; and we are dabblers in the kingdom of God in part because our individual human weaknesses are so strong, and in part—a largely unexamined part—because the kingdom of God has been betrayed by formal Christian teaching for well over fifteen hundred years. The kingdom of God, as a teaching and proclamation integral to the Gospels, has largely disappeared from mainstream Christianity. It's hard to want to live in simple, ecological community when even the *desire* for such community has been disavowed, discredited, and discarded by ecclesiastical teaching and religious institutions for literally centuries on end. Orthodox Christianity has evaded and avoided the kingdom of God, and in that evasion it has betrayed the critical core of Jesus' teaching. Deep inner peace—a meditative life—requires submersion in the kingdom of God.

There may well be practices that facilitate our meditation. Buddhists, for instance, have much to teach us about silence and self-emptying. We should seek and not scorn such teaching.

But at the molten core of our own spiritual heritage there is a teaching, an invitation, to a simple, complex, earthy, highly devout communal life based on servanthood and stewardship. It's in the Gospels, and it is front and center in the Lord's Prayer: Your kingdom come, your will be done, *on Earth* as it is in heaven. It is not a prayer based on spiritual dabbling, nor is it a social vision subservient to civilized mythology. The kingdom of God seeks to *dissolve* predatory civilization by creating a powerful social organism, a spiritual culture of servanthood and stewardship. The kingdom of God is an intensely spiritual form of ecological community. The kingdom of God is Green.

Notes

1. Borg, *God*, 61.

10

An Ethos Spawned and Nourished

It's become a familiar complaint: the Sunday school stories young children are taught are not outgrown, are not retold with increasing complexity, depth, and critical analysis, and therefore spiritual maturation is blocked or thwarted. Those who can't abide such childishness jump ship. Many who remain in the church—at least a disproportionate number—become rather rigid "fundamentalists," with or without quotation marks, people whose spiritual life is boxed in by stories they may be privately embarrassed to believe but afraid, for various reasons, to openly disbelieve. Thus the literal veracity of Bible stories can become a point of contention, especially when fiercely asserted by apparently unembarrassed professional believers, against which "science" is cast (with or without quotation marks) as, at best, an inferior expression of truth. Truth develops opposing camps: "religious truth" versus "scientific truth"—as if there were two worlds instead of one or as if religious truth has only a single, absolute mold. If the complex wisdom of biblical narrative is either abandoned as childish or reified as Absolute Truth, a critical but hard to achieve process of spiritual growth can get sidetracked, scorned, and neglected.

The particular issue, the specific context, I want to put forward as a contrasting concept is "civilization."

In Genesis, the biblical narrative informs us that Cain killed Abel and became a founder of cities. (How, suddenly, there were so many people remains unexplained, and problems of incest are not delved into.) But somehow, from the "fundamentalist" point of view, Cain's fratricide and city building just has to be believed. Just has to, that's all. Because the Bible is God's word, and God's word is true, literally and absolutely. If you can't quite grasp it, just say "It's a mystery" and move on.

Those even roughly familiar with scholarship in regard to prehistory either shake their heads in disbelief or have a good, if bitter, laugh.

"Fundamentalists" assert that there is no such thing as "prehistory," except, perhaps, for the Eternal One's brooding over unformed chaos prior to Creation. Human life started in Eden. Adam and Eve were evicted from the Garden (Genesis 3:14–24) as a consequence of disobedience (eating of the Tree of the Knowledge of Good and Evil) and potential future disobedience (eating of the Tree of Life). Therefore "fundamentalists" have no truck with the scholarship historians are engaged in, specifically in regard to the dawn and origins of civilization. Precivilization rests in Eden, period, and nowhere else.

If Eden is all there is to know, then gatherers, in the late Stone Age, methodically developing horticulture in the direction of cultivated fields, larger yields, increased human population and stable villages, is just atheistic drivel. It's drivel because God's word is true and absolute. The "Stone Age" is a diabolical fantasy. Since nothing (except divine brooding) happened before the six-day Creation, before Eden, there is nothing to be learned (except wicked and deceptive speculation) about human origins or the evolution of human culture that predates the story of Genesis. Absolutely nothing.

I don't think it's possible to overestimate how this mythological amnesia/anesthesia has served, and continues to serve, as a religious moat around the aristocratic castle. Fundamentalism—perhaps paradoxically—shields civilizational myth. Fundamentalism—as was covertly discernible in the politics of Ronald Reagan and overtly visible in the politics of George W. Bush—serves to buttress and protect civilization from critical analysis. That is, the religious refusal to know anything about human history prior to Eden and Cain serves as an immeasurably important protection for the dual assertion that God gives and sustains civilization, and that civilization is an agency (perhaps the key agency) for achieving cultural and even spiritual refinement. If Cain (as the Bible says) is the founder of civilization, and if Cain and his works are under God's protection (as Augustine taught), who would dare be critical of civility?

Is it possible to have dialogue along these lines of confrontation and denial? When "scientific truth" and "religious truth" are mutually exclusive assertions, something has to be in error.

Well, I have no remedy that presumes having cake and eating it, too. Biblical narrative might be rich in spiritual wisdom, but it's not always

An Ethos Spawned and Nourished

literally true. As the astronomy fights several centuries ago should inform us—in June of 1633, Galileo was summoned to Rome, examined by the Inquisition ("under menace of torture," according to the *Encyclopaedia Britannica*), and compelled to recant his heretical Copernican views—it's really quite stupid for people with religious convictions to presume that biblical statements about the nature of reality are always empirically true, even in the face of incontrovertible evidence to the contrary.[1] Even "fundamentalists" now believe that Earth orbits the Sun, and not the other way around.

Prehistory may not be as easily proved as certain aspects of astronomy. It can't be reduced to mathematics or the careful observation of "heavenly bodies" in orbit. It might be possible to tuck Earth-going-around-the-Sun into biblical literalism; but it's not possible to tuck in prehistory. If prehistory is true, then the Garden of Eden story is not literally true, and "fundamentalism" falls apart—and maybe even suffers a nervous breakdown. That is, to break open the rigidities of fundamentalist belief is tantamount to spiritual crisis. It implies an entire reorientation of what's meant by "belief" and "faith." It probably requires a long, painful wrestle with a collapsed or collapsing image of God.

But (in my unscientific estimation) a huge number of secularists and also religious liberals have a parallel fixation to fundamentalism. That fixation is "civilization." I will try to show how these fixations overlap.

---

1. 1. It's not exactly fair to suggest by omission that Protestants were more intellectually responsible than Catholics. In *The World of Copernicus*, on page 94, Angus Armitage points out that although "Catholics showed a dislike of the teachings of Copernicus, we might have supposed that therefore their opponents, the Protestants, would have embraced the new theory eagerly. But actually the Protestants rejected the Copernican system even more decisively than the Catholics. Copernicus had been in his grave for many a long year before the Catholic Church made an official pronouncement that his teachings must be rejected. But even before the publication of his book, Copernicus had been severely criticized by Luther, who, in conversation, used to denounce 'the new astronomer who wants to prove that the Earth goes round, and not the heavens, the Sun, and the Moon; just as if someone sitting in a moving wagon or ship were to suppose that he was at rest, and that the Earth and the trees were moving past him. But that is the way nowadays; whoever wants to be clever must needs produce something of his own, which is bound to be the best since *he* has produced it! The fool will turn the whole science of Astronomy upside down. But, as Holy Writ declares, it was the Sun and not the Earth Joshua commanded to stand still.' Luther's colleague, Philip Melanchthon, also argued against the Copernican theory in a little book on physics which he published some six years after the astronomer's death. Like Luther, he laid especial stress on texts in the Bible which seemed to suggest that the Earth was at rest."

The Kingdom of God Is Green

Well, let's just open the package. Augustine, in *The City of God* (Book V: 21), said: "... we do not attribute the power of giving kingdoms and empires to any save to the true God."[1] (Augustine's remark, it seems to me, is an extension of the Apostle Paul's assertion in Romans 13 that there is no governing authority "except from God.") Now kingdoms and empires are forms of civilization. So the conclusion to this little Augustinian syllogism is that God gives and sustains civilization. We all have (one presumes) reasonably accurate ideas of what "kingdom" or "empire" means in conventional historical understanding. These are political formations, inherently hierarchical in nature, built around and built up on an economic infrastructure based in certain geographical areas.

Prior to the industrial revolution, every civilization rested on the back of undemocratic coerced labor, sometimes outright serfdom or overt slavery. In 1 Samuel, Chapter 8, God has Samuel warn the "elders of Israel" about the "rights of the king," what the price of civilization really is. This is the point in Jewish history when the elders are determined to have a king, a monarchy, so to "be like all the other nations." Samuel, speaking for Yahweh, delivers what may be the most explicit denunciation of civilization in the entire Bible. Here's what Samuel says (10–18):

> All that Yahweh had said Samuel repeated to the people who were asking him for a king. He said, "These will be the rights of the king who is to reign over you. He will take your sons and assign them to his chariotry and cavalry, and they will run in front of his chariot. He will use them as leaders of a thousand and leaders of fifty; he will make them plough his ploughland and harvest his harvest and make his weapons of war and the gear for his chariots. He will also take your daughters as perfumers, cooks and bakers. He will take the best of your fields, of your vineyards and olive groves and give them to his officials. He will tithe your crops and vineyards to provide for his eunuchs and his officials. He will take the best of your menservants and maidservants, of your cattle and your donkeys, and make them work for him. He will tithe your flocks, and you yourselves will become his slaves. When that day comes, you will cry out on account of the king you have chosen for yourselves, but on that day God will not answer you."

One can, by slowly gathering in as spiritual nourishment the sayings and parables of Jesus, come to an astonishing revelation of how starkly and totally in opposition to civilization the kingdom of God is, aided by (e.g., Ephesians 6:12) the warnings against Powers and Principalities,

An Ethos Spawned and Nourished

exemplified by what William Stringfellow calls "the Constantinian Arrangement," an arrangement whose long-term consequences James Carroll methodically exposes in his *Constantine's Sword*.

Stringfellow says (I'm quoting from a little section called "The Constantinian Status Quo," in *A Keeper of the Word*) that by the accommodation of the Constantinian Arrangement,

> . . . signaled by the conversion of the emperor and the establishment of Christianity as the official "religion" of the Roman empire, a comity between church and nation was sponsored that, in various elaborations, still prevails in the twentieth century. The incidents that occasioned the Constantinian Arrangement, as such, are not as significant for contemporary Christians, or for either church or state today, as the ethos spawned and nourished by that comity and the mentality that has been engendered and indoctrinated by it over so long a time. It is, put plainly, an ethos that vests the existence of the church in the preservation of the political status quo. This inbreeds a mentality, affecting virtually all professed Christians, and most citizens whether Christian or not, which regards it as normative for the church's life to be so vested. And that has caused radical confusions in the relations of church and nation, church and state, church and regime. It has encouraged and countenanced stupid allegiance to political authority as if that were service to the church and, a fortiori, to God. Venerable though it be, this accommodation, and the way of conceiving of the juxtaposition of church and political authority that it has inculcated for so very long, accounts more than anything else for the profound secularization of the church in the West and for the inception of Christendom as the worldly embellishment of Christianity.[2]

Stringfellow goes on to say that this arrangement represents the "historical reversal of the precedent established in the apostolic church regarding relations with political authority." He says this "inherited political and theological stance" must be "transcended" and Christian people "emancipated from the indoctrination of the Constantinian mentality."[3]

Well, Constantine may have strong-armed the Church into a forced marriage with Empire; but it was Augustine who wrote the pious homily and gave the enduring blessing and, more critically, provided the rationale and justification for this marriage by pronouncing Empire God's will. So, let's cut through the benedictions and look at what civilization is and risk the sharp displeasure of religious fundamentalists and the ecclesiastical defenders of civilization alike. That is to say, let's look at what happened to the women's fields in the late Stone Age.

Let me provide, immediately, a partial list of sources. This includes many books by Lewis Mumford, including *The City in History* and *The Transformations of Man*; Elise Boulding's *The Underside of History* and *Cultures of Peace*; William Irwin Thompson's *The Time Falling Bodies Take to Light*; Gordon Childe's *What Happened in History*; Warren Johnson's *Muddling toward Frugality*. In a nutshell, here's the summation: women gatherers developed horticulture through a slow but increasingly deliberate process of domestication of wild grains. This led, in the area we now call the Middle East, to small fields, an enlarged and more stable food supply, a shift in diet, an increased human population, and stable villages. Horticulture also enabled the domestication of animals with a somewhat similar diet to humans—that is, cows, sheep, pigs, goats, poultry, donkeys and horses—and to a great many inventions, including the wheel.

All this is not yet civilization. If we are to characterize this precivilized period, it is the golden age of the agrarian village. That is, this productive abundance, this new cultural richness, was not yet systematically expropriated by an armed (male) banditry. At the point where we have armed expropriation, we also have the early city, kingship, a hired or coerced military, and systemic involuntary servitude—that is, massive slavery—exactly as Samuel says.

It is this configuration of aristocracy and involuntary servitude that we have come to know as civilization, whether Babylonian, Egyptian, Greek, Roman, Chinese, Aztec, Incan, or preindustrial European. The locus of this involuntary servitude was the peasant village. That is to say, agricultural abundance was initially a great blessing; but at (and after) that point where an armed and predatory aristocracy imposed itself on the village and villagers, *abundance* was stolen, labor was *compulsory*, and *scarcity* became enforced.

This pattern of "civilized" aristocracy, living off of the expropriated wealth of the peasant village, is the core economic dynamic of all civility up until the industrial revolution. At that point a massive new delusion became common, a delusion persisting to this day, a delusion that parallels and deepens the Constantinian Arrangement that Stringfellow said has caused "radical confusions." If the Constantinian delusion was that the kingdom of God could become or had become the bride of Empire, the industrial delusion was that civilization could become or had become democratized.

An Ethos Spawned and Nourished

Let's try to look more closely into this dynamic.

First, it's a given that civilization up until the industrial revolution (notwithstanding limited forms of aristocratic "democracy," as in Plato's Greece or in Republican Rome) was explicitly undemocratic and overtly authoritarian—that is, based on coercion, expropriation, and involuntary servitude. But yet our present common belief is that modern civilization, at least our American civilization, is "democratic." How and at what point did an inherently undemocratic set of institutions and aristocratic behaviors suddenly become "democratic"? How and when did this happen? Did the undemocratic simply disappear or evaporate at the time of the American Revolution? Did undemocratic behavior ("lording it over") suddenly transform itself into "democratic" forms? If the Civil War ended slavery, did it also eliminate the sense of racial superiority on which slavery was justified? Was the relentless dispossession of Native American peoples and cultures the moral right of white-skinned people proclaiming Manifest Destiny? Is it totally incidental that the dispossession of the European peasantry resulted in a disoriented and disorienting migration that helped achieve the dispossession of indigenous, noncivilized cultures on this continent?

There are several layers here, and the layers interpenetrate. First, the industrial revolution did not just put mechanized power into the production process as it pertains to factory labor; it also put mechanical power into food production. It wheeled the machine into the garden. This latter process, combined with the systematic eviction of peasants from the commons (I'm thinking here primarily of England, where the industrial revolution really took off), forced evicted, unemployed, and often homeless rural folks to either work for starvation wages in the hated factories or starve. An enlarged labor force working for wages in turn led to a greater market demand for food products and therefore to an increasingly commercial food production. The commercial market thus helped to destroy household self-provisioning and cooperative subsistence. The commons and the peasantry were simultaneously forced out of existence. The industrial consequences of the Constantinian Arrangement simply smashed Adam's original vocation (from the biblical narrative) and eradicated (from the natural history narrative) the ancient tradition of Neolithic gardeners. The "surplus" the traditional aristocracy had previously extorted from the peasant village was now increasingly extorted by a commercializing civilization directly from nature, from Creation, via an always intensifying agritechnology. That is to say, agriculture became increasingly mechanized and commercialized as factory production consolidated. All this was done under an enlarging capitalist ideology. The old landed aristocracy came to an end, and a new "democratic" aristocracy of industrial wealth arose.

In this political and cultural turmoil, especially at the end of the eighteenth and up into the early nineteenth century, the upheaval of traditional civilization resulted in limited democratic openings, most specifically in procedures for enlarged participation of "commoners" in election to public office. In the second decade of the twentieth century, so-called communism (state capitalism) took command in Russia, with its subsequent Stalinist attempt to create a workers' civilization out of the czarist state.

What's notable about both capitalist and communist ideologies is that neither repudiated the traditional civilized practice of land confiscation or of extracting wealth out of agriculture. In fact, both found ways to *accelerate* this extraction in order to construct massively industrialized super-states armed with the deadliest weapons the world has ever seen: deadly enough to bring mammalian life to an end the world over.

Stalin collectivized Russian (Soviet) agriculture, so better to *collect* its production and to wring labor out of the countryside. Western capitalism achieved the same end more slowly but more efficiently: in the United States, less than two percent of the population are now engaged in food production; and everywhere the food economy, at the level of the small to medium family farm, is characterized by actual or near depression.

If this is "democracy," then democracy is inherently turned against agriculture. If this is "civilization," then civilization is explicitly weighted against the countryside, against Creation. Well, our current system is certainly not democratic, except in a superficial sense; but it most definitely is civilized. We might even say it is or has become hypercivilized.

We have seen (it is overwhelmingly part of the historical record) that civilization "got its living" by extorting the wealth of the peasant village—its wealth, its production, and its labor. The so-called "democracy" of contemporary society "gets its living" in exactly the same way, although its actual functioning is hidden from us by layer upon layer of economistic psychobabble about "free market" behavior and the like—the "science" of exploitation—even as the underlying dynamics in the heated "illegal immigration" debate rarely touch on the core issues: the destructive impact of "free trade" policies south of the border, and the economic demand in the United States for cheap labor in food production.

An explicit aristocracy of globalized wealth accumulation has arisen, based on unbelievably low reward for small-scale food producers and agricultural workers, sweatshop labor, highly rationalized resource extraction of all sorts, limited deindustrialization in unionized industries, Third World debt entanglements, an utterly massive and deadly military to enforce economic control, and the studied ridiculing and ignoring of impending ecological and toxicological disasters. Nevertheless, virtually all "responsible" public commentators continue to celebrate civilization as if it were our real religion (which it is), and those who would call civilization into question as the most massive form of institutional sin in human history are seen as either lunatics, ranters, comedians, or harmless court jesters.

Unable and unwilling to face up to or come to terms with the inherently predatory and undemocratic nature of civilization, even as global society lurches from one deadly disaster to another, we should nevertheless take warning of how deeply entrenched are our illusions and delusions regarding "civility," "civilized values," and other such highly politicized and even spiritualized terms. It is, I believe, entirely within the biblical understanding of the kingdom of God to suggest that civilization is the name of our true secular religion, and that, in contrast, "kingdom" teaching is so weak and atrophied that even fervently "believing" Christians are panicked by the thought of civilization as historically evil, inherently violent and predatory. Fundamentalists have had their minds wired by Augustine of Hippo.

It's my conviction that Jesus invited us to live by cooperative subsistence in Creation, in spiritual community, renouncing the world (i.e., civilized predation and its ideologies), practicing personally *and* politically as fully as we are able both servanthood and stewardship. We are to love one another even onto death. We are to engage in healing and in sharing, striving only to be the least. We are to live in Creation as gentle stewards.

The emerging, unfolding global crises are the poisoned fruit of civility, civilized values, civilization, and our worship of its golden calf. If Augustine was right, Samuel got it wrong. But if Samuel got it right, then Augustine bears an enormous responsibility for justifying the "Constantinian Arrangement." The Augustinian justification is yet another "Sunday school story" we urgently need to outgrow.

Notes

1. Augustine, *City*, 174.
2. Stringfellow, *Keeper*, 259.
3. Stringfellow, *Keeper*, 261.

11

Servanthood, Stewardship, and a Restraint on Vice

I RARELY BUY A new book. My wife Susanna and I are folk musicians who play, mostly, in nursing homes. These mundane facts converged when, on a recent Friday afternoon, after a musical engagement, we stopped at our favorite thrift store. There I found, for fifty-six cents, a 1970 paperback entitled *The New Left and Christian Radicalism* by Church of the Brethren scholar Arthur Gish.

A few days later, Susanna came home with a brand new paperback, a big one, called *All Saints: Daily Reflections on Saints, Prophets, and Witnesses for Our Time*, by former Catholic Worker Robert Ellsberg. Ellsberg's book is divided into 365 entries, a saint a day, and I went right to July 25, with its entry for Baptist preacher and theologian Walter Rauschenbusch, who died of a brain tumor on July 25, 1918.

And then there is a third book, *Solutions to Violence*, edited by Colman McCarthy, that Susanna bought directly from McCarthy when he spoke in northern Wisconsin in the fall of 2001, in the aftermath of September 11. Wausau peace activist David Kast brought McCarthy to Wisconsin, and two women from Merrill, Irene Mehlos and Laurie Kaufmann, organized a Peace Study, using McCarthy's book, that continues to meet almost every week.

Solutions to Violence is divided into sixteen chapters, each a collection of short pieces by various writers, organized around a theme. Last night, for instance, our group finished Chapter Seven, all articles in regard to the death penalty; and this led us, in a roundabout way, into a discussion of the "two kingdoms."

And that takes us back to Arthur Gish.

The New Left and Christian Radicalism is an explicit effort to contrast and compare the "new" Left of the '60s with the turmoil of sixteenth-century Anabaptist radicalism in parts of (mostly) German-speaking Europe.[1] The tension, dissonance, and outright conflict between community and establishment, between revolution from the bottom up versus reform from the top down, between radical and liberal, informs much of Gish's book. But the core dynamic lies in "Anabaptism: A Sixteenth-Century Analogy," Gish's second chapter. There are really two parts to this core: the doctrine of "two kingdoms" and the kingdom of God. (As I hope the reader will see, the kingdom of God is far too robust, far too unpredictably alive, to passively behave itself as the long-suffering, otherworldly, and sanctimonious part of a mere *doctrine* of "two kingdoms.")

The Anabaptist doctrine of the state, says Gish, "rests on a two-kingdom dualism which makes a sharp distinction between the kingdom of Christ and the kingdom of this world."[1] (The "kingdom of Christ" is, for me, already a little shaky, for I am relying on Gospel language, particularly Luke's, that, in words attributed to Jesus, is explicit as "kingdom of *God*." But—perhaps—this may be a distinction without a difference.) On the following page, Gish says of Anabaptist conviction that "Those who wish to follow [Jesus] cannot be in both kingdoms at the same time."[2] And then, one more page into the chapter, he says Anabaptists "saw the need for people to live now as if the kingdom of God were already here. This was at the heart of their value system."[3]

Jump ahead ten pages and we find that "Anabaptists saw the use of force and revenge as inherent in the role of government."[4] Drop back a page: "Since [Anabaptists] accepted the authorities as ordained by God, they believed in obedience to the government."[5]

I. Roland Bainton, in *Here I Stand*, deals only briefly with Anabaptism. But it's clear (pages 267, 281, and 378) that both Luther and Zwingli (the Swiss reformer) turned on the Anabaptists with wrath. Luther and his fellow reformer Melanchthon "were quite as much convinced as was the church of the inquisition that the truth of God can be known, and being known lays supreme obligations upon mankind to preserve it unsullied. The Anabaptists"—with their wish to restore primitive Christianity, their holding to the Sermon on the Mount, their renunciation of oaths or the use of the sword, their pacifism, religious communism, simplicity and temperance—"were regarded as the corrupters of souls." Bainton also says that the "agrarian complexion of the Anabaptist movement is not by any means wholly the result of the Peasant War but much more of the persecution which could more readily purge the cities than the farms.... Luther's stand was contributory to the alienation of the peasants."

The Kingdom of God Is Green

OK. There are two kingdoms. God ordained both. One is based on love and sharing, healing and forgiveness. The other is based on force and revenge, self-interest and retaliation. You can't be in both at the same time.

Since God ordained both, does this mean that God is schizophrenic and needs either a good psychiatrist, a politically shrewd advisor, or some suitable pharmacology? In Chapter Four, Gish tells us that "This split between the personal and the social has been one of the biggest tragedies of Western ethics."[6] So how does this tragic split between the personal and the social shake out in relation to the two-kingdom doctrine?

> Because of this split [Gish writes] we reason that certain acts committed by societies and institutions are not as evil as the same acts committed by individuals. As Paul Furfey (in *The Respectable Murderers: Social Evil and Christian Conscience*) has so well put it,
>
> "It is an infinitely tragic fact that the greatest crimes of history are committed with the cooperation or at least with the passive consent of the solid citizens who constitute the stable backbone of the community. The sporadic crimes that soil the front pages, the daily robberies, assaults, rapes, and murders are the work of individuals and small gangs. They are committed by manifest criminals whom the community despises and punishes. But the great evils, the persecutions, the unjust wars of conquest, the mass slaughters of the innocent, the exploitations of whole social classes—these crimes are committed by the organized community under the leadership of respectable citizens."
>
> The church has reinforced this view through relegating the gospel to private and interpersonal relationships. This has fit well with the development of capitalism. Modern industrial society considers the role of religion to be in the subjective, personal, and individualistic aspects of life. It is careful to keep religion out of those areas which would be threatened by the Christian perspective. It sees faith as having no significance for social and bureaucratic relationships. Thus God is not the God of society, the world, or history. He becomes a privatized idol.[7]

Arthur Gish does not push this analysis to its logical conclusion. That conclusion, in my estimation, is that the doctrine of two kingdoms is bogus, untrue, pernicious, and a great wounding constraint on the kingdom of God. I realize that this conclusion puts one at odds with a lot of doctrine that is (or can be construed as) Bible-based, from "chosen people" to Romans 13:1–7, to preoccupation with "private" salvation, a lot of which

Servanthood, Stewardship, and a Restraint on Vice

is formulated in the books that follow the first three Gospels in the New Testament. Yet the task, it seems to me, for every person who calls herself or himself a Christian, is not to go to sleep in the comfortable nest of one's chosen sect, or even to accept Paul's writings as having equal theological weight or spiritual insight as the Gospels, but to labor mightily to find Jesus amid, behind, and below all the hugely accumulated theological clutter of twenty centuries, including all the legalistic paperwork attendant on the fourth-century Christian merger with Empire. If God is God of society, the world, and history, then the *doctrine* of two kingdoms, with its separate but more-than-equal status for the state, is a blind—perhaps not so blind—lawyers' trick, a loophole in God's love, based far more on Plato than on Samuel, all of which serves to neuter and "domesticate" the kingdom of God.

The kingdom of God *is not* the "spiritual" half of a God-ordained spiritual/secular two-kingdom duality. The kingdom of God is the whole and entire loaf, not just half a loaf, much less a mere crusty heel. The Creator is most certainly complex and subtle beyond all our most vivid imaginings; but we have it on the assurance of Jesus that Spirit is not psychotic. We know from early modern astronomers that the Earth goes around the Sun. We know from astronomers, geologists, biologists, historians, archaeologists and anthropologists that not only is the universe (and our very own Earth) immensely old, but human beings have existed and had complex culture long before there ever was such a thing as a state, long, long, long before the rise of that great economic squeeze machine called civilization. To give the civilized state equal footing, equal billing, with the kingdom of God is a staggering pomposity. It is antithetical to everything Jesus was about. That the state conspired to kill him might be taken as something of a clue.

The current globalization of civilization shows just how incredibly devious the doctrine of two kingdoms is, for it asserts that in the kingdom of the state the kingdom of God has no license to practice and needs to stick to harmless sacraments and otherworldly travel planning. The two-kingdom doctrine opens Creation wide to complete utopian control by the civilized state, while the kingdom of God is relegated to its tax-free shelter on the corner of Third and Main, so long as it keeps its pious nose out of "politics." The state gets to play by its own set of rules based on greed, retaliation, force, and wrath, while both the state's politicians and the church's compliant theologians invoke God to justify this corrupt arrangement.[II]

II. Well, it's not just "compliant theologians." Arthur Gish, for instance, on page 56,

The Kingdom of God Is Green

Let's revisit Robert Ellsberg's two-page glimpse into the insights of Walter Rauschenbusch. In 1891, Ellsberg says, Rauschenbusch "experienced a theological breakthrough." This breakthrough was his "rediscovery of the biblical symbol of the kingdom of God, the central message Jesus had proclaimed." This central message

> . . . had been either forgotten or obscured in Christian history. The kingdom was identified with heaven, with individual salvation, or with the church. While it was truly concerned with all these things, [Rauschenbusch] argued, the kingdom was ultimately something wider and more radical. Above all it represented the constant tension between the status quo and the world as God intended it.[8]

I don't think the kingdom of God is "above all" a mere *tension*, for the maintenance of tension is itself a capitulation to the continuation of the status quo. The kingdom of God is a transformational process intending to *overcome* the status quo and, thus, *overcome* the tension (untruth) that protects the status quo. So it's not incidental that Rauschenbusch's "great contribution," says Ellsberg, was the "articulation of an understanding of 'social sin,' a term that has only resurfaced in the language of the church."[9] So what's the connection between "social sin" and the kingdom of God?

"Social sin" is collective *institutional* sin, that form and magnitude of sin the state is the enforcer of, the protector of, and even the locus of in civilized history. The basic business of the state has always been, historically, the expansion and protection of empire. It is built on greed, theft, and oppression. It is the plaything of aristocracy. In our world, institutional sin is primarily commercial greed linked to the military machine of the civilized state. It is the great repository of the status quo. Yet institutional sin is not limited to the state. The church, too, when it insists on maintaining a false doctrine or a false understanding, is playing in the toxic waters of institutional sin. The construct of two kingdoms is such a false doctrine.

---

says "The Anabaptists had a clear doctrine of the state. There was little doubt among most of them that the state (*Obrigkeit*) is ordained by God. The task of the state was considered to be closely connected with its origin. It had existed from the beginning of creation with God as the ruler. After the fall of man, however, this office was given to man. The state, then, must be understood in light of man's sin." The doctrine of two kingdoms, in other words, depends both on a monarchical image of God and on a literalist reading of the creation account in Genesis. Without these underlying foundations, the doctrine of the two kingdoms is shown to be made of sawdust and glue, a metaphysical construct based on false premises.

Servanthood, Stewardship, and a Restraint on Vice

The rise of democracy in the last couple of centuries has, it is true, caused the state to assume a horizontal dimension without letting go its vertical dimension. By "vertical" I mean those governing instrumentalities that oversee and protect the concentration of wealth and the consolidation of power in the hands of a few. By "horizontal" I mean the use of government to thoughtfully improve and prudently enlarge the cultural sensitivity and economic well-being of all persons, with an eye always toward ecological health and social stability. (Vernon Louis Parrington, in *The Colonial Mind: 1620–1800*, quotes the early American diplomat Joel Barlow: "If government be founded on the vices of mankind, its business is to restrain those vices in all, rather than to foster them in a few."[10] Or, as Arthur Gish also says, "government is for man's protection, for the preservation of order. Its purpose [is] to protect the innocent and weak from evildoers, to maintain peace and order."[11])

In Marxist eschatology, the end of theft and war is the "withered" state. From a Christian perspective in which the state (and, more importantly, humanity as a whole) is slowly but steadily overcome with what we might call the divine biology of the kingdom of God, it's the theft and war attributes of the state that will wither. The kingdom of God is also a kind of globalization: it seeks the well-being of all people, the ecological health of all places, and it desires all of it to be in tender touch. This requires coordination, decision-making, a modest amount of management and prudent planning.

The two-kingdoms doctrine is pernicious in that it limits the kingdom of God to that "spiritualized" domain fenced off by Paul and Augustine. It is inherently Manichean and fundamentalist in that it insists the two kingdoms are forever separate, spirit and matter, light and dark, ordained by God. This gives carte blanche, unconditional terms and full discretionary power to the state, and it renders the church at best a yipping moral puppy, but, when petted and well-fed, a comfortable lap dog capable of growling as only pet doggies can.

If we say and believe that the Creator also created our earthly speck of cosmic debris—and this may be foundational to our faith, although I think we need to be more humble with our assertions about what God supposedly has done or intends—then how is it we deviate and detour from this basic conviction by abandoning Creation to wrath, to revenge and to evil, and then smugly assert that this abandonment is the will of God? It's really pretty amazing. Awe and reverence in the midst of mystery—and what is

57

this life but mystery?—seems to me a far more wholesome response than proprietary presumption and calculated self-seeking.

If the state is the arm of God's wrath, and if the state is free to kill for reasons of empire protection, retribution, and revenge, then it has to be said that God (despite Paul's remark, in the opening chapter of Hebrews, that Jesus is the "perfect copy" of God's nature) always keeps a little torture in reserve for the dirty bastards who've got it coming. When the state executes a prisoner, murders a designated "enemy," keeps "enemy combatants" locked away offshore and without trial, or initiates a war of subversion or retaliation (a bow here to Colman McCarthy and the local Peace Study gathering), the state invariably will say either God made me do it or God allows me to do it. Civilization, say the two-kingdom apologists, needs this kind of moral slack, for men—they mean other men—are evil creatures in need of our restraining power.

The heart of the matter is a direct parallel to Galileo's run-in with the church in the seventeenth century. The church had a powerful, theological vested interest in an Earth-centered cosmology. But the astronomers were right and the church was wrong. "Faith" had seriously misplaced its attention and allegiance. In its "protection" of God (and/or biblical truth), the church instead revealed its stupidity.

Likewise, the church (and the state) continues to have a theological vested interest in the doctrine of the two kingdoms. By long tradition, the church is conditioned to evade the kingdom of God. It's in love with its own spiritual laziness, which is the flip side of its otherworldly and metaphysical preoccupations. (One would be hard pressed to justify the two-kingdom doctrine from the synoptic Gospels. Paul is another matter. Augustine is explicitly and overtly another matter.)

The astronomers were right. The church was wrong. Similarly, the scholarship on prehistory, the origins of civilization, and the theft and war history of civilization simply explode the doctrine of two kingdoms. The state is not the arm of God's wrath specifically designed to protect the weak from the strong. The state is the civilized control center whose enforcer arms are theft and war. The doctrine of two kingdoms cannot be postulated from the sayings or teachings of Jesus.

If you are a fundamentalist who "believes" absolutely in "the Bible"— Luther's *sola scriptura* or some variant—then you are obliged, I suppose, to be as persuaded by Paul as you are by Jesus. If the Bible is, from end to end, the inerrant Word of God, then all its teachings and admonitions have equal status: Joshua ready for genocide at Jericho has as much moral

Servanthood, Stewardship, and a Restraint on Vice

credibility as Jesus saving the accused woman from stoning. And while Paul can truly be, here and there, our teacher, only Jesus is our Teacher. To make Paul's assertions (or Augustine's) equal to those of Jesus may be good Manichean bibliolatry, but it is repulsive to faith. If we deify the Bible, then we are stuck in bibliolatry. The doctrine of two kingdoms, like the Christian mainstream emphasis on otherworldly salvation, is largely misplaced allegiance. Neither otherworldly salvation nor the two kingdoms is the central point. With Jesus, the point is our capacity and willingness to enter the kingdom of God—and live there. What makes us think we will merit eternal salvation if we categorically refuse to embrace the kingdom of God on Earth?

The Jesus I know was executed by the state. Did God kill Jesus with "His" other arm? Is this all an allegorical game? Does God's lawyer offer an insanity plea? Is this a case of divine forgetfulness in which one hand truly didn't know what the other hand was doing?

God may be (in the words of Arthur Gish) a "privatized idol" for those Christians who have not yet broken through their salvation-saturated bibliolatry; but God is also a public idol used to justify the state and the state's predatory actions. This is not to say that human beings can be expected to live without a form (or forms) of self-governance, or even that those forms of self-governance can exist without the invocation of law. It is to say that the state that serves the interests of theft and war, of wrath and limitless human appetite, is finally an institution of evil, Powers and Principalities, and that the kinds of self-governance that can and will arise from the kingdom of God are, first and foremost, instruments of compassion and caring, servanthood and stewardship, and, secondarily, a restraint on the vices of all.

If theft and war are the wicked arms of Civilization, servanthood and stewardship are the healing hands of the kingdom of God. As Robert Ellsberg quotes from Walter Rauschenbusch: "When a man has once seen that in the Gospels, he can never unsee it again."[12]

Notes

1. Gish, *New*, 57.
2. Gish, *New*, 58.
3. Gish, *New*, 59.
4. Gish, *New*, 69.

5. Gish, *New*, 68.
6. Gish, *New*, 96.
7. Gish, *New*, 96–97.
8. Ellsberg, *All*, 317.
9. Ellsberg, *All*, 318.
10. Parrington, *Colonial*, 386.
11. Gish, *New*, 56.
12. Ellsberg, *All*, 317.

12

Clinging to Dead Ideas

IF IT SEEMS EXCESSIVE and perhaps even reckless to say that civilization is the locus of our real religion, our actual devotion to self-seeking and retaliation, allow me to shortly quote a paragraph from the eighth chapter of Kevin Phillips' *Wealth and Democracy*. I think Phillips points right into the depths of this abyss, even though his language is couched in a sophisticated sort of wary sarcasm: he *looks* in, with a jaunty, devil-may-care use of terminology, but he doesn't really *go* in.

The chapter is entitled "Wealth, Money, Money-Culture Ethics, and Corruption," and it has little fingers of analysis, reaching as far back as the Renaissance, revealing how the "seven deadly sins of the Middle Ages—pride, gluttony, avarice and prodigality, lust, sloth, anger and envy—were converted into the driving values of the Renaissance era," and how "yesteryear's private sins and vices—individual compulsions to self-interest, avarice, luxury, and pride—can and do reemerge from time to time as commercial and civic virtue, indeed props of unusual national success."[1]

Phillips casually employs terms like "vice-into-civic-virtue theology," "free-market core theology," and "tax cut theology."[2] But nowhere does he truly follow up on or explore the dreadfully scary implications of these terms, except—here's the paragraph—to say this about religion:

> Religion, too, has had its voice in the conservative economic chorus. From colonial times, preachers had often obliged. During the Gilded Age the Reverend Henry Ward Beecher lauded business for fat speaking fees, and Baptist minister Russell Conwell became a millionaire from the appeal of "Acres of Diamonds," his sermon to large and prosperous crowds that getting

rich was a noble aspiration. The twenties had Bruce Barton's story of Jesus as the world's greatest salesman. President Calvin Coolidge even confused religion and economics when he observed that "The man who builds a factory builds a temple. The man who works there worships there." Kindred voices in the nineties were Paul Zane Pilzer, author of *God Wants You to Be Rich*, Catherine Ponder of the Unity Church Worldwide who penned *Dare to Prosper*, and Deepak Chopra, author of *The Seven Spiritual Laws of Success*.[3]

If we have paid even modest attention to politics over the last thirty years or so, we cannot help but recognize how closely "conservative" Christianity has allied itself with "conservative" politics, and even how they have reinforced and to some degree accelerated each other's growth. Well-funded foundations and "think tanks"—many of which have a foot in each "conservative" camp—have enormously aided this process of alignment by projecting a high public visibility and by achieving "talking head" status in the media, especially television, which also, in turn, has come under the control of huge corporations that have been deeply involved in, and benefited from, the very processes Kevin Phillips dissects. It's a tight loop. (Perhaps the modern linkage of "conservative" Christianity to "conservative" politics should remind us of how and why "orthodox" Christianity aligned itself with the Roman Empire in the fourth century. The monarchical model or image of God really does have long-term political consequences.)

The "conservative" stance is, largely, unintellectual, highly moralistic, easily folded within a racist positioning, prone to intense individualism (of which "family values" is largely an extension), reflexively anxious about homosexuality and therefore inclined to identify homosexuality as a sin of major proportion and consequence, patriotic with little apparent knowledge of the depth and complexities of American empire machinations (but what is known is intensively colored by Christian triumphalism, Manifest Destiny, anticommunism, a deep hostility toward "paganism" or cultural "backwardness," simple racism), and full automatic deferment to assertive male authority that either invokes or smoothly symbolizes all these values. That these values essentially saturate much of the church is simple sociology. That they are largely and collectively the antithesis of the Gospel, that they congeal politically as the value core of globalized empire, that they reflexively support militarism, capital punishment, capitalism and economic growth, that they are easily brought into alignment against any or all of the world's other great religions (even identifying them, when politically expedient, as evil), that they are full and unhesitating defenders and promoters of civility, civilized values, and civilization—all this brings us to the terrible and terrifying conclusion that Christianity is, in many ways, the antithesis of the Gospel, the enemy of Jesus, and even the Anti-Christ.

Such an outrageous thought brings us to that bad boy of Christian theology, Thomas J. J. Altizer, whose 1965 book, *The Gospel of Christian Atheism*, was at the core of the late '60s controversy regarding "the death of God." This is a book I've only now read. *The Gospel of Christian Atheism* is not a light-weight read, and I will be thinking about it for a long time to come. I don't know that I accept, or ever can accept, Altizer's assertion that, by becoming incarnate in Jesus, God has "ceased to exist in his original mode."[4] That's way too big an assertion to get my mind around. I sense it asks me to make a decision about God or come to a conclusion about God that is beyond my capacity for anything resembling certainty. I am only a human creature. My knowledge doesn't include precise familiarity with God's birth date or obituary.[I]

The importance of Altizer's book, however, is not really contained in its assertion that God has "ceased to exist in his original mode." The significance lies in an obliquely related matter: what happens to Christianity when it has fully bought into and fully aligned itself with the accumulated and cumulative institutional sin of civilization.

I. If Thomas J. J. Altizer tells us that "God has ceased to exist in his original mode," then Albert Nolan, in his *Jesus Before Christianity*, tells us that "*God has changed*" (Nolan's emphasis, page 95), "God has come down from the heavenly throne" (page 97), "God has now revealed God as the God of compassion" (page 102), and God is the "one who has now changed and relented of God's former purposes in order to be totally compassionate toward humankind—all humankind" (page 167).

Perhaps this sort of thinking is simply beyond my pay grade, but I believe such pronouncements about God represent sincere attempts to go beyond conventional depictions or orthodox images of God while simultaneously trying to keep God intact. (Nolan also says on page 167 that we must not "abolish the Old Testament" or "reject the God of Abraham, Isaac and Jacob." But "if we accept Jesus as divine, we must reinterpret the Old Testament from Jesus' point of view.") In my estimation, these assertions reveal an inability or unwillingness to step beyond biblical convention. Radical change is seen as crucial, change that necessitates a radically altered depiction of a wrathful Father God; so therefore God is dead or has voluntarily humbled Himself for our edification.

Joachim of Floris—as we shall see in "The Age of the Daughter"—provided us with a way out of this box. Once we take constructs of the divine out of an exclusively metaphysical ether and ground them in history (and this is exactly what Joachim did), we have a path for a far more comprehensive understanding that no longer needs theological gymnastics explaining God's psychotherapy or mortification. This grounding may create ambiguity about the divinity of Jesus, but that's a human psychological problem, not a cosmic problem needing First Responder emergency resuscitation.

The Kingdom of God Is Green

So allow me to bring forward from Altizer what I consider of primary importance. It involves, of course, the kingdom of God:

> We know that the proclamation of both Jesus and the earliest Palestinian churches revolved about the announcement of the glad tidings or the gospel of the dawning of the Kingdom of God. But thus far neither the theologian nor the Biblical scholar has been able to appropriate the eschatological symbol of the Kingdom of God in such a manner as to make it meaningful to the modern consciousness without thereby sacrificing its original historical meaning. It is scarcely questionable, however, that this symbol originally pointed to the final consummation of a dynamic process of the transcendent's becoming immanent: of a distant, a majestic, and a sovereign Lord breaking into time and space in such a way as to transfigure and renew all things whatsoever, thereby abolishing the old cosmos of the original creation, and likewise bringing to an end all that law and religion which had thus far been established in history. The very form of Christianity's original apocalyptic proclamation rests upon an expectation that the actualization of the Kingdom of God will make present not the almighty Creator, Lawgiver, and Judge, but rather a wholly new epiphany of the deity, an epiphany annihilating all that distance separating the creature from the Creator. Despite Paul's conviction that the victory which Christ won over the powers of sin and darkness had annulled the old Israel and initiated the annihilation of the old creation, to say nothing of his assurance that God will be all in all, both Paul and the early Church were unable fully or decisively to negate the religious forms of the old history, or to surmount their bondage to the transcendent and primordial epiphany of God. Consequently, early Christianity was unable either to negate religion or to absorb and fully assimilate an apocalyptic faith, with the result that it progressively became estranged from its own initial proclamation.[5]

> Whether or not we choose to so understand the original Christian gospel of the dawning of the Kingdom of God, it is clear that the radical Christian affirms that God has died in Christ, and that the death of God is a final and irrevocable event. All too obviously, however, we cannot discover a clear and decisive witness to the meaning of this event in either the Bible or the orthodox teachings and visions of Christianity. . . . A contemporary faith that opens itself to the actuality of the death of God in our history as the historical realization of the dawning of the

Kingdom of God can know the spiritual emptiness of our time as the consequence in human experience of God's self-annihilation in Christ, even while recovering in a new and universal form the apocalyptic faith of the primitive Christian.[6]

Once again we have detected a Christian religious reversal of God's act in Christ: for a faith that isolates the sacred events of Christ's passion from the profane actuality of human experience must inevitably enclose Christ within a distant and alien form and refuse his presence in the immediacy of our existence. Every Christian attempt to create an unbridgeable chasm between sacred history and human history gives witness to a refusal of the Incarnation and a betrayal of the forward-moving process of salvation.... All such religious claims not only attempt to solidify and freeze the life and movement of the divine process, but they foreclose the possibility of the enlargement and evolution of faith, and ruthlessly set the believer against the presence of Christ in an increasingly profane history, thereby alienating the Christian from the actuality of his own time. The radical Christian calls upon his hearer to open himself to the fullness of our history, not with the illusory belief that our history is identical with the history that Jesus lived, but rather with the conviction that the death of God which has dawned so fully in our history is a movement into the total body of humanity of God's original death in Christ. Once we grasp the radical Christian truth that a radically profane history is the inevitable consummation of an actual movement of the sacred into the profane, then we can be liberated from every preincarnate form of Spirit, and accept our destiny as an occasion for the realization in the immediacy of experience of the self-emptying or self-annihilation of the transcendent and primordial God in the passion and death of Christ.[7]

It is not without significance that the modern artist has given himself so fully to envisioning evil and nothingness, or has been so deeply bound to visions of Satan, of chaos, and of emptiness; for the artist cannot escape the reality of his time by fleeing to an earlier moment of history. A new epiphany of Antichrist is drawing everything into itself, as its ever dawning totality transfigures all experience, unveiling the emptiness of Hell in every human hand and face.

Yet if the Christian can but name our Hell as Antichrist, then we shall know that its power has been broken and it can pose no ultimate threat to us. When all evil and nothingness pass into the faceless epiphany of a total Antichrist, then the ultimate ground of chaos will be dissolved, every inherent sanction for all alien and compelling demands will be removed, and every opposing other will stand revealed in a lifeless and vacuous form. The reign of the Antichrist is the consummation of all oppressive power.... It is precisely because an epiphany of Antichrist abolishes the transcendent source of evil and nothingness by embodying a primordial chaos in the actuality of history that it is a redemptive epiphany, an epiphany unveiling the full reality of alienation and repression, thereby preparing the way for their ultimate reversal.... [W]e must nevertheless be prepared to open ourselves to the anguish and terror of experience as an expression of the atoning process of redemption.[8]

From the point of view of radical Christianity, the original heresy was the identification of the Church as the body of Christ. When the Church is known as the body of Christ, and the Church is further conceived as a distinct and particular institution or organism existing within but nevertheless apart from the world, then the body of Christ must inevitably be distinguished from and even opposed to the body of humanity.... Thus a forward movement in this sense is finally the expression of the will to power, an all too human regression to an inhuman or prehuman state, which necessarily entails a reversal of the true humanity of Jesus. Once the Church had claimed to be the body of Christ, it had already set upon the imperialistic path of conquering the world, of bringing the life and movement of the world into submission to the inhuman authority and power of an infinitely distant Creator and Judge.

But by identifying the Church's Christ as a reversal of the incarnate Christ, a reversal effected by a backward movement to the now emptied preincarnate epiphanies of God, the radical Christian points the way to the presence of the living Christ in the actuality and fullness of history. It is precisely because the orthodox image of Christ is an image of lordship and power that it is a reversal of a kenotic Christ. The mere fact that the Christ of Christian orthodoxy is an exalted and transcendent Lord is a sufficient sign to the radical Christian that Christianity has reversed the movement of the Incarnation. Simply by clinging to the religious image of transcendent power, the Church has resisted the self-negating movement of Christ and foreclosed the possibility of its own witness to the forward movement of the divine process. Consequently, the radical Christian maintains that it is the Church's regressive religious belief in God which impels it to betray the present and the kenotic reality of Christ. So long as the Church is grounded in the worship of a sovereign and transcendent Lord, and submits in its life and witness to that infinite distance separating the creature and the Creator, it must continue to reverse the movement of the Spirit who progressively becomes actualized as flesh, thereby silencing the life and speech of the Incarnate Word.[9]

If Christ is truly present and real to us in a wholly incarnate epiphany, then the one principle that can direct our search for his presence is the negative principle that he can no longer be clearly or decisively manifest in any of his previous forms or images. All established Christian authority has now been shattered or broken: the Bible may well embody a revelation of the Word but we have long since lost any certain or even clear means of interpreting its meaning as revelation; the Church in its liturgies, creeds, and confessions may well embody an epiphany of Christ, but that epiphany is distant from us, and it cannot speak to our contemporary experience. Even the language that the Christian once employed in speaking of Christ has become archaic and empty, and we could search in vain for a traditional Christian language and symbolism in contemporary art and thinking.[10]

With the death of the Christian God, every transcendent ground is removed from all consciousness and experience, and humanity is hurled into a new and absolute immanence. Our chaos becomes manifest as a uniquely modern chaos when it is ever more comprehensively present in response to the emptying of the transcendent realm, as its darkness fills every pocket of light, and night falls throughout the whole gamut of experience. Now an ultimate choice is thrust upon every man, as he can either turn back in horror at our chaos by engaging in a final No-saying, or he can turn forward and meet our darkness by means of an ultimate Yes-saying, a total affirmation of our actual and immediate existence. Such an acceptance and affirmation is possible only if man will give all of the energy which he once directed to a transcendent beyond to the immediate moment, thus releasing every source of energy so as to effect a total engagement with the actual present before him.[11]

Every nostalgic yearning for innocence, all dependence upon a sovereign other, and every attachment to a transcendent beyond, stand here revealed as flights from the world, as assaults upon life and energy, and as reversals of the full embodiment of love. The Christian who chooses the ancient image of Christ as the Son of God, or who is bound to an epiphany of Christ in a long-distant past, must refuse the Christ who is actually present in our flesh. He wagers upon a purely religious image of Christ even at the price of forfeiting the actuality of our time and history. But the radical Christian wagers upon the Christ who is totally profane. He bets upon the Christ who is the totality of the moment before us, the Christ who draws us into the fullness of life and the world. Finally, radical faith calls us to give ourselves totally to the world, to affirm the fullness and the immediacy of the present moment as the life and the energy of Christ. Thus, ultimately the wager of the radical Christian is simply a wager upon the full and actual presence of the Christ who is a totally incarnate love.[12]

Whether all this is theologically true in some absolute and empirical sense is, frankly, beyond me. (Was God transcendent before the time of Jesus? But became immanent by breaking into time and space and thus has "ceased to exist in his original mode"? I am inclined to believe that mysterious Spirit constitutes the real "monotheism" and that other, more rationally graspable constructions—including Mother, Father, Son, and Daughter—may be helpful as images and even contain elements of Spirit's radiance appropriate to their respective epochal designations; but they are not to be taken literally as divine Persons. They are as fictitious as corporate "personhood.") Yet the spiritual correlation to our actual and existing religious circumstance is astonishingly (and frighteningly) apt. When Christians fiercely claim to be eternally "saved" because they "believe in" Jesus, but are in their actual lives at least as fiercely for capitalism, free trade, globalization, military prowess, unrestrained energy production and commodity consumption, capital punishment, and suspicious of (if not openly antagonistic toward) ecological prudence, racial equality and sexual reconciliation, universal healthcare, mass transit, a more fully democratic United Nations, and sustained compassionate dialogue with the rest of the world (including the world's other great religions), then the church is very much part of an idolatrous, imperialistic effort set on conquering the world in behalf of a "transcendent," otherworldly ideal. Implicitly if not explicitly it has renounced servanthood and is contemptuous of stewardship. Doesn't the larger condition of Christianity, especially "conservative" Christianity, confirm Altizer's disturbing assertions—and confirm them with growing force?

These imperialisms congeal and coalesce in the New Rome that America has become. For what is the *deification* of self-interest, greed and consumption, and the *theology* of tax cuts, except demonic empire parasites feeding on the lacerated and decaying body of a dying Creation and perhaps even on the putrid carcass of a dead God? The church's betrayal of Jesus, the Gospel, and the kingdom of God is now reaching its culmination in the deification of a spurious god, an un-Jesus, an Anti-Christ, whose name is Empire Civilization, shamelessly asserting its heroic red, white, and blue righteousness with unrivaled weapons of annihilation and belittling if not simply denying that its economic utopianism is in process of shredding Earth's ecology.

In the September 2002 issue of *The Progressive*, in "The Progressive Interview" conducted by David Barsamian, Kevin Phillips again invokes religious imagery without explanation or exploration. In response to a question, he says "Conservatives worship at the altar of supply-side tax management, which is tax cuts. They think tax cuts cure everything but warts."[13] But nowhere does this conversation enter the terrain of foundational ethics. The spotlight turned on the concentration of wealth is brilliant but so narrowly focused. Never do we get anything substantial or visionary about what democracy means or could mean, much less about the spiritual origins of humanitarian ethics. Phillips, by his self-imposed narrowness, reduces himself to a brilliant, carping gadfly.

Meanwhile, the breaking of this empire lies just ahead. Our task is to bear witness to the Gospel, to hope, pray, and work for the kingdom of God to arise not so much in empire's place (for this would be only a duplication of the imperialistic path of conquering the world) as to redeem with immanent grace the space empire once transcendently hogged. To be ready for this, to keep our lamps lit and not fall asleep, requires that we struggle hard in prayer, study, dialogue, and reflection in order to cleanse our hearts and minds of false conviction, even if those convictions are traditional and dear to us. To be stupidly unready because we cling to dead ideas would be a tragedy of immense proportions.

Notes

1. Phillips, *Wealth*, 334, 329.
2. Phillips, *Wealth*, 335.
3. Phillips, *Wealth*, 338.
4. Altizer, *Gospel*, 69.
5. Altizer, *Gospel*, 105–6.

6. Altizer, *Gospel*, 107–8.
7. Altizer, *Gospel*, 108–9.
8. Altizer, *Gospel*, 121–22.
9. Altizer, *Gospel*, 132–33.
10. Altizer, *Gospel*, 137.
11. Altizer, *Gospel*, 150–51.
12. Altizer, *Gospel*, 156–57.
13. Barsamian, "Progressive," 36.

13

The Engine of Disaster

BOTH FUNDAMENTALISM AND MAINSTREAM Christianity contain major obstacles to deepening and extending our understanding of the connections between spirituality and social justice. In the past, the dreadful significance of these obstacles and obstructions was not recognized or appreciated, except by a dismissible few. But the global crises, contained like restless golems within the morphogenesis of civilization and now bursting forth in the contemporary world—not so much as blistered externalities as erupting internalities—force us to recognize and appreciate these obstructions, and to break through them into a larger and more spiritually faithful understanding.

In the case of fundamentalism, its literal six-day creationism radically obstructs any understanding of prehistory or of how prehistorical social developments impact present-day circumstances. In the case of mainstream Christianity, its ongoing weddedness to what William Stringfellow called the "Constantinian Arrangement" provides a major barrier to examining and exposing the predatory core of civilization, both in its origins and in its maintenance of power to this day.

As human beings we are each and every one of us occasionally threatened by truth, perhaps more than we are aware of, certainly more than we care to acknowledge or admit. We do not live in anything resembling perfect transparency. Most of us, it seems, not only have a private or even secret self, but we often act in ways that protect that private, secret self. Much of this protectiveness involves our need to be seen as normal, as fitting in; and that, of course, requires that we participate in upholding the norms. But as people who call ourselves Christian, we are under instruction—an instruction lubricated by grace—to constantly struggle through our blockages and get past our resistances to truth, even if this involves going against the norm. Furthermore, our faith in the boundless subtle intelligence and limitless overflowing compassion of Spirit teaches us not to fear the "natural history" discoveries of science, even though our inherited religious or mythological worldview might be bruised or broken. We can confidently trust that a bigger and more magnificent worldview will be forthcoming. There's always room for a new norm.

It is my understanding that the universe is truly ancient. Physicists and astronomers may or may not have all the facts in hand, as it were; but the natural sciences, including geology and archaeology, have exploded the conventional Christian understanding of a literal six-day creation roughly six thousand years ago. Those who persist in this worldview cling to an idolatry of literalism, demanding that Spirit (and all of us) conform to the outline of a discredited story, even if that story was taught to us as God's infallible Word.

On the other hand, those who fall within the contemporary denominations captivated by the Constantinian Arrangement, those who follow Augustine's assertion in *The City of God* (Book V:21) that kingdoms are given and sustained by God, rather than Samuel's warning to the elders of Israel regarding the "rights of the king" (1 Samuel 8:11–17), have neutered the kingdom of God by aligning the church, permanently, with empire. In the instance of fundamentalism, the sustained hostility toward prehistory renders the appearance of civilization magical: Cain suddenly became a "founder of cities" without precedent or context. The careful science of prehistorical reconstruction is simply dismissed as delusion, a trick of the Devil. Natural scientists are, thereby, tools of Satan.

Augustine also took old stories rather literally—and embellished them with (one presumes Manichean) indulgences of Angels of Light and Angels of Darkness—making of this world a kind of Platonic cave allegory. Heavenly salvation was Augustine's open goal. This earthly life was only to be endured. Therefore empire, as the organizer of civilized structure and enforcer of civilized order, was not only welcomed, it was the will of God, a sort of transitional, but intensely hierarchical, social glue meant to hold the human world together before extraction to heaven or descent into hell. If heaven was the otherworldly carrot held aloft by the church, civilization controlled the earthly stick.

With Constantine and Augustine, the marriage of Christian church and civilized state was announced, celebrated, and sanctified. The kingdom of God, as taught, described, and practiced by Jesus, became narrowly transmogrified into a far removed eschatological concept, the Second Coming of Christ, arriving only at the End of Time. The kingdom of God as Gospel yeast, as slow but steady transformer of human community on Earth, was suppressed in favor of formal eschatology and ecclesiastical hierarchy. Soul transport was substituted for life transformation. Heaven heaved the kingdom of God overboard. The institutional church baked its Jesus yeast in the oven of doctrinal orthodoxy. Or, in Altizer's finer language, the early church was unable to surmount its "bondage to the transcendent and primordial epiphany of God."

Needless to say, Spirit's yeast is more than any institution can freeze, kill, or bake in a doctrinal oven. It is not finally constrained by any bondage, though it may well be retarded by assertive human stupidity. (What would become of Spirit in the event of all-out nuclear war is simply beyond my comprehension.) But it's true that there has been, and is, a primary or mainstream understanding of what Christianity is, of what Jesus represents. Fundamentalism and mainstream Christianity, though mutually hostile in many particulars, overlap in a way that has served to block, obstruct, and retard our spiritual understanding of several critical matters intimately tied to the kingdom of God. This blockage has extremely important implications and ramifications in regard to the rapidly enlarging and intensifying ecological crises. I will try to briefly sketch a portion of the blockage and its significance.

By denying prehistory, fundamentalists obstruct and prevent an understanding of the origins of agriculture and the formation of civilization. The long and ancient history of gatherers and hunters is said never to have existed. By denying prehistory, we are forbidden to examine the dynamics inherent in the slow development of horticulture, the stable village, steady growth in human population, the domestication of animals, and, finally, agriculture. Civilization, in the fundamentalist worldview, springs full-blown from the head of Cain. Civilization, in mainstream Augustinian ideology, is God's will and therefore under sacred guidance and sacred protection. Cain was the founder of cities. Augustine blesses and sanctifies those cities. End of story.

Not the end of story.

Contemporary ecological crises—which many scientists warn us are growing at an alarming and unprecedented rate, the consequence of a progressively "advanced" civilized standard of living imposed simultaneously on Creation and on all the world's cultures—force us to read the story backward, to find where the story went wrong. (As a note here, we can also point out that fundamentalists have two primary responses to global warming and climate change. The first is denial. The second is that it's the End of the World. Either and both fit the idolatry of biblical literalism. Meanwhile, mainstream Christians, addicted to affluent civilization in its full industrialization, say the problem is too many people and outmoded technology. Therefore the "solution" is population control and technical fix. This stance basically accepts civilized sanctity and is inherently Augustinian.)

We need to take this story back before the Bible narrative. The Bible does not here help us see the critical formative dynamics of civilization, though it may help instruct us in how we respond to them.

Ten thousand years ago, more or less, horticulture was in the very slow process of being created, primarily (I'm not up-to-date on the latest research) in the area of present-day Iraq, between the Tigris and Euphrates Rivers, part of the so-called Fertile Crescent. Historians say gatherers—the women—were the core practitioners of this development. Their domestication of wild foods, especially the grains, triggered a food abundance that led, eventually, to the domestication of a variety of animals, to stable villages, and to new inventions like the wheel. What's critical to grasp is that this new agrarian and village abundance is *not* civilization.

Civilization is a system of aristocratic economic extraction and political control based on the violent expropriation of primary production. Agriculture resulted in village abundance. Civilization extorted that abundance and created village scarcity. Civilization did two totally novel things: it institutionalized the military and it institutionalized slavery. The connecting link is perpetual taxation of the many for the benefit of the few. This is hardly a new insight. Read 1 Samuel, Chapter 8. Read what Jesus says about "worldly" authority.[I]

All civilizations have, in various ways and degrees, followed this pattern of aristocrat and peasant. A few men rule with swords, with law, with institutional prerogative, and with a forged divine blessing. The bulk of the common people, overawed, labor under compulsion for the rulers and secondarily for their own subsistence.[II]

I. See also Marcus Borg's compact summation, pages 134 through 136, in *The God We Never Knew*. I quote some of Borg's summation in "The Perfectly Camouflaged Temple," pages 26 and 27, in my *Green Politics Is Eutopian*.

II. Reinhold Niebuhr in his *Faith and Politics*, pages 83 and 84, says "religious reverence for majesty is a more potent force of subjection than fear of power" and that "The most potent government is one in which the functions of priest and soldier are combined. Stated in other terms, reverence for majesty and fear of force are the two

The Engine of Disaster

Because we have all been taught that civilization is a spiritual process of moral refinement raising us from primitivity, barbarism, and paganism, and because we lack, almost totally, a critical analysis of how civilization came into being or how it lives by predation and expropriation, we tend to wallow in moral-cliché deference, gravitating toward apologetic and well-intentioned social services, ethically and historically incapacitated by our unwillingness to grasp the deeper dynamics of social and ecological justice—meek social janitors, humbly cleaning up the institutional messes, never plain-spoken prophets who identify civilization as the multimillennial Powers-and-Principalities world criminal that it is. (Of course Jesus healed the sick. But he also confronted the authorities and the system they represented. Of these two tendencies—healing and confrontation—we can safely say Jesus was not murdered for his skill as a healer.)

The late, eminent world historian Lewis Mumford, in an essay called "Utopia, the City, and the Machine," said that civilization has its origins in utopian aggression—an impulse, not unlike Augustine's, to be rid of this world with its natural limitations, and to build, by coercion, the City of God on Earth. Those origins—and also their aggressions—are with us to this day.

The condition of the present world—thousands of nuclear weapons in apocalyptic readiness, species extinctions, elimination of languages and cultures, depletion of rain forests and fisheries, global pollution, space weapons, a "cultural" saturation with compulsory schooling and electronic entertainment, the destruction of agriculture and the globalization of agribusiness, a military-industrial prison complex, new epidemic diseases, etc.—tells us that something is tragically flawed in the story we've been compelled to live by. If the story was true, civilization's globalization would be in process of creating a true paradise on Earth. Civilization promotes itself (and has promoted itself for thousands of years) as our savior from primitivity, coarseness, and the base material world. Everything that is elegant, refined, noble, of enduring cultural and uplifting spiritual value is said to be *inside* civility. Civilization stands tall against backwardness and terrorism in a way exactly parallel to how institutional Christianity has suppressed "paganism" and "false belief." The globalization of civilization should therefore be resulting in a finer, cleaner, and gentler world, when in fact the opposite is the case. Weapons of Mass Destruction. Mutually Assured Destruction. Global Warming. Climate Change. End Times.

---

most effective motives of obedience to government." Niebuhr goes on to say, however, that "The claim of divine majesty on the part of a state and nation always involves fraud."

The Kingdom of God Is Green

Both fundamentalism and Constantinian Christianity work to achieve and sustain this parallel fixation: the former, through willful ignorance, choosing not to know the origins of civilization; the latter, through Augustinian ideology, asserting civilization as God's will. In the fundamentalist worldview, our life on Earth is only a brief spiritual test that, depending on the rightness or wrongness of our belief, results in inclusion or exclusion in heavenly eternity. In the Augustinian/Constantinian worldview (*The City of God*, Book XIX:13) civilization imposes a necessary order on an inherently corrupt and unruly world and is the earthly form of God's rule. Civilization is God's police force. "Conservative" fundamentalism simply doesn't know, and doesn't seem to care to know, what civilization is or where it came from. "Liberal" mainstream Christianity glosses over civilization's utopian predation because civilization, in principle, has been taught as God's will, and because, in lived reality, centuries of false understanding have resulted in a nearly total acquiescence to and identification with a progressively "advancing" standard of living. Civilization provides our real identity. We're all "middle class" now.

Fundamentalism is historically brain dead. Mainstream Christian liberalism drifts along on the current of civilized globalization, critically anesthetized. This is why we can be approaching—reaching—ecological disasters on a global scale and, simultaneously, observe these disasters with such disregard, detachment, and disinterest. The bibliolatry of fundamentalism and the Manicheanism of Augustinian eschatology combine to make us spiritually numb. In the stance of fundamentalism, there will be rapturous rescue from End Times disaster; in the stance of mainstream liberal Christianity, technofix and attention to overpopulation will see us through.

That civilization is itself the engine of disaster is an assertion unacceptable to both fundamentalists and mainstreamers, but even more unacceptable to the liberal mainstream. When Constantine forced Christianity into marriage with Empire, Augustine pronounced the union God's will. Christianity was blessed by civility, and civilization magically broke free of its alleged "paganism" by becoming "Christianized."

To break out of denial (for fundamentalists) and out of false affirmation (for mainstream Christians) amounts to a very serious spiritual crisis, a crisis with profound ecological, cultural, and political consequences. Or, again, in the complex terminology of Thomas Altizer, the "actualization of the Kingdom of God will make present not the almighty Creator, Lawgiver, and Judge, but . . . a wholly new epiphany of the deity, an epiphany annihilating all that distance separating the creature from the Creator."

Or is it precisely this anxiety-reducing distance we crave, a distance that preserves what we believe to be normal, a distance that protects us from a new epiphany, a distance that permits us to assert our dominion of Earth under the aegis of civilized governance, a distance that leaves our precious egos undisturbed and essentially intact?

The real safety net is, as always, the mysterious net-working of Spirit. More specifically, the safety net is the noncivilized kingdom of God that Jesus saw, taught, pointed toward, and enacted. He invited us to enter this kingdom, to *live* in it as stewards and servants in community, to live simply and trustfully, to renounce "lording it over," to wash one another's feet.

I'm not saying that Jesus didn't have anything to teach about the next world. It's not that salvation (in the sense of some after-death life) is necessarily false. (We'll all know—or not know—soon enough.) It's that self-centered preoccupation with either otherworldly salvation or civilizational sanctity atrophies and falsifies the meaning of the kingdom of God on Earth. To live in the kingdom of God on Earth ("Your kingdom come" is a key part of the only prayer Jesus is recorded as teaching) is our mandate. Our clear tasks are to be stewards and servants, and that means simplicity and sharing. (If eternity is all-embracing and all-encompassing time, then life on Earth is also life in eternity.) But unfold kingdom teaching from within the history and conduct of civilization and we find civilization not only a colossal idolatry, but also the single most concentrated and complex formation of institutional sin in human history.

The kingdom of God is the opposite of civilization. All the power, glory, aggression and contempt that is endemic to civilization—its ceremonial pomp, official self-importance, vicious militarism, economic predation, and sanctimonious self-justification—the kingdom of God abhors and declines, which, in part, is what the temptations of Jesus in the wilderness teach us (Matthew 4:1–11 and Luke 4:1–13). Jesus says no to food magic, circus tricks, and political domination. Jesus chooses life in full trusting dependence on Spirit's invisible strength, mercy, and purpose. He does not choose magic, power, or showmanship. This "kingdom" is unlike any civilized kingdom that has ever existed. This "king" is a servant without pretension to civilized power.

The emerging global crises point toward the actual limitation and the eventual breakdown of civilization. The question for us is not what we can or should do to "save" civilization. Samuel was right and Augustine was wrong. Our question is—What does it is mean to live in the kingdom of God? What must we do to enter? The death of civilization provides a great opening for the kingdom of God. Our job is to pray Your kingdom come as if we meant it—and then mean it.

14

Voting for Jesus

FOR ALL OUR PRETENSIONS to liberation from conventional expectation, we still desire a basic orderliness. We like a beginning, a middle, and an end. Call it the Tao of Contentment. Writers have the (essentially invisible) freedom, when recognizing a weak or underdeveloped spot in narrative sequence, to revisit the past, as it were, and do what is impossible in real life: remedy in "real time" a previously enacted deficiency.

The task at hand is to try, especially by robbing some carefully crafted intellectual eggs out of John Dominic Crossan's scholarly nest, to firm up some tenuous linkages between ancient peasant experience and what Jesus (or at least the Gospel writers) called the kingdom of God. And since Dom Crossan's *The Historical Jesus: The Life of a Mediterranean Jewish Peasant* has a lot of the right stuff by which to build that peasant/kingdom linkage, I'm going to indulge in something I've, in the main, been trying to minimize—quotation.

We will start, however, not with Dom Crossan, but with Marcus Borg. Borg's essay is entitled "The Palestinian Background for a Life of Jesus," and it comes from a little book, published by the Biblical Archaeology Society, called *The Search for Jesus: Modern Scholarship Looks at the Gospels*, edited by Hershel Shanks. Borg spends most of his time on what he calls "five cultural dynamics"—colonial, cosmopolitan, peasant, purity, and patriarchal—dynamics operative in Israel in the first century C.E.[1] In brief, Israel was under the harsh and brutal *colonial* domination of Rome; Galilee (the city of Sepphoris in particular) was far more *cosmopolitan* than we previously imagined; Jesus simply ignored the *purity* prescriptions of

Voting for Jesus

the dominant Jewish culture; and he repeatedly violated the *patriarchal* presumptions that saturated male consciousness. I slid deliberately over *peasant*. Here's Professor Borg on that dynamic:

> A third characteristic of the social world of Jesus is that it was a peasant society. By this I do not simply mean that there were a lot of peasants, although there were. Rather, peasant society is a shorthand phrase for a particular type of society, namely "pre-industrial agrarian society" as described by Gerhard Lenski [in his *Power and Privilege: A Theory of Social Stratification*]. These societies are known widely throughout the premodern world.
>
> The defining characteristic of a pre-industrial agrarian society is that it's a two-class society. On the one hand, there are urban ruling elites, and on the other hand, there are rural peasants. The rural peasants typically comprise approximately 90 percent of the population. To flesh out that grand concept just a bit, the urban ruling elites consist of five groups: the ruler, the governing class, the retainers (retainers are basically employees of the ruler and the governing class); the well-to-do merchants; and the upper echelon of the priesthood. The ruler and governing class are about one percent of the population and typically receive about half of the income. The elites together (ten percent of the population) typically receive two-thirds of the income. The rural peasants include small landholders as well as sharecroppers, day laborers, unclean and degraded classes and expendables.
>
> There's a huge gulf between these two classes. Peasant societies are marked by sharp social and economic inequalities. There is no middle class. To try to illustrate that with two contrasting diagrams, all of us are familiar with the pyramid diagram of modern societies—a fairly small upper class, a larger middle class and an even larger lower class. A peasant society would not be diagrammed as a pyramid. The best analogy I can think of is one of those old-fashioned oilcans with a broad bottom and a long narrow spout coming up out of it. The vast majority of people are represented by that broad bottom and the urban ruling elites by the needlelike spout rising vertically from the base.
>
> Where do the urban ruling elites (not just in first-century Jewish Palestine but generally in societies like this) get their wealth? They don't manufacture anything. They don't produce anything. They don't grow anything. I'm not even sure they provide any services. They get their wealth, of course, from the peasants, and they get it in two forms—rent for land and

> taxation. Peasant societies are thus economically oppressive and exploitative.[1]
>
> This awareness illuminates the Gospels and what the Gospels say about Jesus in a number of ways. I'll mention just a couple for illustrative purposes. When Jesus speaks about his message being "good news to the poor" or when he says "blessed are the poor," it's pretty clear, I think, that he's talking about real poor people. This is not a metaphor. He is talking about the oppressed group in a peasant society.
>
> The teaching of Jesus also includes a number of indictments. The indictments are not of society as a whole but of the elites. Jesus' primary social conflict was with the elites. This is illuminating when we think about the causes of the death of Jesus. In all likelihood, a combination of Roman authority and a narrow circle of the Jewish ruling elites was responsible for his arrest and execution. Very importantly, rather than Jesus being rejected, arrested and executed by "the Jews" or the Jewish people, the final and fatal conflict was with urban ruling elites who, rather than representing the Jewish people, were in fact oppressors of most Jewish people.[2]

The peasantry is the class, in a generic sense, that elite prerogative, cultural bias, spiritual idolatry, and sustained economic predation simply ate alive in the aftermath of the industrial revolution. The subsequent consolidation of both wealth and power—huge wealth, unimaginable power—in the hands of a transnational elite, essentially a new "post-aristocratic" aristocracy, now threatens to exterminate this generic peasant class the world over. This is the global victory of civilization we are generally encouraged to celebrate—the overcoming of rustic backwardness.

Here are questions that no one seems to want to ask: Has Jesus' alliance with the peasantry been rendered irrelevant and made moot by industrial agribusiness? Should we be grateful that "stewardship" is now in the scientifically competent hands of chemical companies and genetic engineering firms? Is it true that the new scientific/technological methodologies for wealth extraction and product fabrication have rendered

---

1. Note that even here, with a deeply sensitive and sympathetic scholar, the semantic convention obscures the reality. Is it *peasant* society that is oppressive and exploitative? The answer is obviously no. It is a *civilized system* that holds peasant society in bondage. But we are never permitted a negative appraisal of civility, even when its brutal reality stares us in the face. Brian McLaren offers a parallel illustration on page 260 of *Everything Must Change*: "Perhaps, in the interests of sustainability, we should speak less of an *environmental crisis* and speak more of an *overconsumption crisis*. That way, we'd focus our attention on the source of the problem, not its victim." Exactly so.

civilization "democratic"? Haven't the subsistence, equalitarian peasant teachings of Jesus simply been smashed by urban-industrial commodity abundance and "progressive" political ideology? Is it true that a "narrow circle" of ruling elites, having both destroyed peasant culture and flooded mass society with the idolatrous propaganda of civility and the technological commodities of utopia, has crucified the peasantry just as, at an earlier and more localized incident, it crucified a certain Mediterranean Jewish peasant? These, it seems to me, are some of the questions we really need to agonize over. Others are: What is it Jesus meant by "kingdom of God"? Does the term have any relevance whatsoever in a world where the organizational, technologized City of Man has seemingly made irrelevant not only the Augustinian City of God but also the visionary, kingless "kingdom" of the Gospels?

The Historical Jesus, by John Dominic Crossan, is a wonderfully witty and lively book. It's also a heavily referenced scholarly tome, loaded with quotations from social science texts, anthropology, history, economics, and, of course, the Bible and its various sources. There's no way I can pretend to a scholarly competence in review of Crossan's work, even when considering his subtitle: "The Life of a Mediterranean Jewish Peasant." I do claim, as a consequence of my rural upbringing and my haphazard studies of the last thirty-five years, to know something about peasants and the cultural value of the peasantry. I also claim to be a Christian with a particular sort of peasant slant to my convictions. (To assert oneself as a peasant Christian, or Christian peasant, is, in postmodern America, at least amusing.)

I say "peasant" even though my upbringing was post-World War II, Midwestern American. I was raised on a small, first-generation, diversified dairy farm in north-central Wisconsin. We lived by a rich combination of self-provisioning, woods work, and milk check—and so did every other family in the neighborhood. These families were in the last wave of small-farm homesteading in America, the last gasp of the Jeffersonian vision.[II] From a "scholarly" point of view, I got to see American agrarian folk culture in localized, miniature formation—telescoped, as it were, into a long generation, before the postmodern economy snapped the telescope shut and left us with supermarkets, superhighways, consolidated school systems, color television, computers, Super Bowl football, and The Last Remaining SuperPower.

II. I explore this theme in "An Unseemly Eulogy for My Father," the last chapter in *Polemics and Provocations*, pages 150 through 173.

The Kingdom of God Is Green

John Dominic Crossan was part of something called the Jesus Seminar, a series of scholarly gatherings convened by Robert Funk of the Westar Institute that worked to discuss, and even vote on, the reputed sayings of Jesus—or, as John Dominic Crossan says, to "try and establish some scholarly consensus on the historical Jesus." (A red bead meant Yes, I think Jesus said this. A pink bead meant Jesus Maybe said this. A grey bead meant Perhaps not. A black bead meant No, he didn't say this. What a bright and colorful thing! A bunch of Christian Reds voting for a Pinko Jesus!) Aggravating "wit" aside, the quest for the historical Jesus proceeds by the careful peeling away of theological addenda, trying as hard as possible to see the real man behind all the layered theological costumes. This process makes lots of people uneasy. Some it just makes mad; that biblical scholars would dare to unravel scriptural threads seems unspeakably presumptuous if not downright evil.

The question, then, comes down to what we really mean by reverence for God. Do we treat our *religion* as if it were sacred? Our doctrines? The Bible itself? Or is our passion for Spirit at once so intense and so serene that we know we are at liberty to explore every nook and cranny of our spiritual inheritance?

Much of the Christianity I am familiar with—both Protestant and Catholic—is simply mired in an obtuse "belief system" that has made an idolatry of "believing" certain doctrinal formulations to be absolutely true: that by a somewhat mysterious mental process of convincing yourself that specific statements about Jesus or God or the Trinity or the Bible or the church are absolutely accurate and unconditionally true, your "salvation"—your acceptable entry into heaven after death—is assured. Jesus is a divine salvation figure whom I therefore must mentally cling to with all possible conviction. This is the ultimate effort of Positive Thinking. In this dynamic, it is not Jesus' teachings that are the fuel of my faith, but his magical image or, more simply, a pervasive saturation in creed.[III]

The more I ponder this situation (it's one I've struggled with since childhood), the more convinced I am that clinging to Jesus as a divine salvation figure serves not only as a substitute/evasion for a difficult transformation to a life of servanthood and stewardship, it also devolves into its opposite. That is, it becomes mythological idolatry, fearful and self-obsessed image worship and creedal superstition.

III. It may well be that lots of church attendees are not at all sure what they believe; liturgy therefore becomes the pre-established and presumably secure hat rack on which to hang one's Sunday headgear for an hour or so each week.

Just about everything in John Dominic Crossan's scholarly quest for the historical Jesus underscores the conclusion I've more or less already come to: that Jesus was a flaming radical who, were he to wander into a conventional church service today, would be reduced to fits of astonished laughter, sobs of stunned frustration, and a wild, angry overturning of cushioned pews. Let me cite a few particulars from *The Historical Jesus*:

> Historical Jesus research is becoming something of a scholarly bad joke. There were always historians who said it could not be done because of historical problems. There were always theologians who said it should not be done because of theological objections. And there were always scholars who said the former when they meant the latter.[3]

> I am concerned, not with an unattainable objectivity, but with an attainable honesty.[4]

> Rome and Italy had learned what other great empires also learn. An imperial heartland can export its violence elsewhere and call it law and order, can even call it peace.[5]

> Aristocrats do not consider themselves "above" peasants but beyond them somewhere, in another world.[6]

> Bandits are peasants grasping for a hold on that most unpeasant phenomenon, power.[7]

> That social ambiguity of banditry as poised between unpower and power has been pushed to an even deeper level by Brent Shaw in discussing its legal ambiguity between right and might specifically in the Roman Empire. The tone is beautifully set by his article's twin epigraphs. The first is from Augustine's *City of God* 4.4: "Remove justice and what are states but gangs of bandits on a large scale? And what are bandit gangs but kingdoms in miniature?" To political ambiguity is then added religious ambiguity, as the next quotation is from Luke 23:33–43, where Jesus is crucified between two not, as we sometimes say, thieves but bandits.[8]

> What was a bandit but an emperor on the make, what was an emperor but a bandit on the throne?[9]

> Rural banditry holds up to agrarian empire its own unpainted face, its own unvarnished soul.[10]

The Kingdom of God Is Green

> The Galilean peasants might not have been able to imagine a new social order, but they could well imagine a world with certain administrative centers razed to the ground.[11]

> The peasantry, in other words, were being pushed below the subsistence level that they considered their normal lot in life. Recall, however, that we are talking about relative or perceived deprivation. The Palestinian peasantry, like all peasants before and after them, lived as close to bare subsistence as those who controlled them could calculate. . . . They had always been double taxed, for foreign empire and indigenous temple; they had always suffered droughts and famines. . . . It will not do for us to say, therefore, even with some degree of social and political morality, that they revolted because the Romans were imperialists, taxed them harshly, or kept them at subsistence level. For there to have been a *perceived* deprivation, the peasants must have been pushed below the subsistence level, not into poverty, which was "normal," but into indigence and destitution.[12]

> [I]t was obviously possible for the first Christian generations to debate whether Jesus was for or against the ritual laws of Judaism. His position must have been, as it were, unclear. I propose . . . that he did not care enough about such ritual laws either to attack or to acknowledge them. He ignored them, but that, of course, was to subvert them at a most fundamental level. Later, however, some followers could say that, since he did not attack them, he must have accepted them. Others, contrariwise, could say that, since he did not follow them, he must have been against them. Open commensality profoundly negates distinctions and hierarchies between female and male, poor and rich, Gentile and Jew. It does so, indeed, at a level that would offend the ritual laws of *any* civilized society. That was precisely its challenge.
>
> Another, and even more important postscript: is all of this simply projecting a contemporary democratic idealism anachronistically back onto the performance of the historical Jesus? I emphasize most strongly, for now and the rest of this book, that such egalitarianism stems not only from peasant Judaism but, even more deeply, from peasant society as such.[13]

> A note on terminology. I am not particularly happy with the word *kingdom*. It is not only androcentric—that, at least, admits its historical bias—it is also primarily local or, at least, is readily so interpreted. But what we are actually talking about is power and rule, a state much more than a place, or, if you will, a place only because of a state. And, lest one ambiguity replace another,

state means a way of life or mode of being, not nation or empire. The basic question is this: how does human power exercise its rule, and how, in contrast, does divine power exercise its rule? The kingdom of God is people under divine rule, and that, as ideal, transcends and judges all human rule. The focus of discussion is not on kings but on rulers, not on kingdom but on power, not on place but on state.[14]

The idea of the rich in the Kingdom is not only quite impossible, it is rather hilarious, like getting a camel through the eye of a needle.[15]

"It is hard," as Douglas Oakman rightly concludes, "to escape the conclusion that Jesus deliberately likens the rule of God to a weed." . . .

The point, in other words, is not just that the mustard plant starts as a proverbially small seed and grows into a shrub of three or four feet, or even higher, it is that it tends to take over where it is not wanted, that it tends to get out of control, and that it tends to attract birds within cultivated areas where they are not particularly desired. And that, said Jesus, was what the Kingdom was like: not like the mighty cedar of Lebanon and not quite like a common weed, like a pungent shrub with dangerous takeover properties. Something you would want in only small and carefully controlled doses—if you could control it.[16]

The third image is just as bad or even worse. The parable of . . . *The Leaven*. . . involves leaven, a woman, and the act of hiding, all of which have negative connotations in their original social matrix. . . .

The essential point is "that leaven in the ancient world was a symbol of moral corruption," according to Brandon Scott, since it was "made by taking a piece of bread and storing it in a damp, dark place until mold forms. The bread rots and decays . . . modern yeast . . . is domesticated." Furthermore, "in Israel there is an equation that leaven is the unholy everyday, and unleaven the holy, the sacred, the feast." . . . Once again, we are confronted with an image of the Kingdom that is immediately shocking and provocative. And it is compounded by the fact that, again from Scott, "woman as a symbolic structure was associated in Judaism, as in other Mediterranean cultures, with the unclean, the religiously impure. The male was the symbol for purity." Furthermore, "the figurative use of hiding to describe the mixing of leaven and flour is otherwise unattested in Greek or Hebrew." . . . With mustard and darnel, then, stands another and triply shocking image for the Kingdom: a woman hiding leaven in her dough. It's there, it's normal, it's necessary, but society has a problem with it.[17]

The sapiential Kingdom looks to the present rather than the future and imagines how one could live here and now within an already or always available divine dominion. One enters that Kingdom by wisdom or goodness, by virtue, justice, or freedom. It is a style of life for now rather than a hope of life for the future. This is therefore an ethical Kingdom, but it must be absolutely insisted that it could be just as eschatological as was the apocalyptic Kingdom.... My proposal is that when we cross apocalyptic and sapiential with scribes and peasants, it becomes necessary to locate Jesus in the quadrant formed by sapiential and peasant. What was described by his parables and aphorisms as a here and now Kingdom of the nobodies and the destitute, of mustard, darnel, and leaven, is precisely a Kingdom performed rather than just proclaimed.[18]

Jesus' Kingdom of nobodies and undesirables in the here and now of this world was surely a radically egalitarian one, and, as such, it rendered sexual and social, political and religious distinctions completely irrelevant and anachronistic.[19]

Jesus sets parents against children and wife against husband, sets, in other words, the Kingdom against the Mediterranean. But not just against the Mediterranean alone.[20]

[W]e are not just dealing with almsgiving but with a shared table, with commensality. The missionaries do not carry a bag because they do not beg for alms or food or clothing or anything else. They share a miracle and a Kingdom, and they receive in return a table and a house. Here, I think, is the heart of the original Jesus movement, a shared egalitarianism of spiritual and material resources. I emphasize this as strongly as possible, and I insist that its materiality and spirituality, it facticity and symbolism cannot be separated. The mission we are talking about is not, like Paul's, a dramatic thrust along major trade routes to urban centers hundreds of miles apart. Yet it concerns the longest journey in the Greco-Roman world, maybe in any world, the step across the threshold of a peasant stranger's home.[21]

Commensality was, rather, a strategy for building or rebuilding peasant community on radically different principles from those of honor and shame, patronage and clientage. It was based on an egalitarian sharing of spiritual and material power at the most grass-roots level.[22]

I emphasize, once more, that the historical Jesus had both an ideal vision and a social program.[23]

There is, however, one further step to be taken. Most of Jesus' first followers would know about but seldom have experienced being served at table by slaves. The male followers would think more experientially of females as preparers and servers of the family food. Jesus took on himself the role not only of servant but of female. Not only *servile* but *female hosting* is symbolized by those four verbs. Far from reclining and being served, Jesus himself serves the meal, serves, like any housewife, the same meal to all including himself. Later, of course, and quite legitimately, it would happen that just as the female both serves food and becomes food, so Jesus would both have served food here below and would become food hereafter (Bynum). But long before Jesus was host, he was hostess.[24]

His strategy, implicitly for himself and explicitly for his followers, was the combination of *free healing and common eating*, a religious and economic egalitarianism that negated alike and at once the hierarchical and patronal normalcies of Jewish religion and Roman power. And, lest he himself be interpreted as simply the new broker of a new God, he moved on constantly, settling down neither at Nazareth nor Capernaum. He was neither broker nor mediator but, somewhat paradoxically, the announcer that neither should exist between humanity and divinity or between humanity and itself. Miracle and parable, healing and eating were calculated to force individuals into unmediated physical and spiritual contact with God and unmediated physical and spiritual contact with one another. He announced, in other words, the brokerless kingdom of God.[25]

The challenge that deeply faithful scholars like Marcus Borg and Dom Crossan pose for us is not simply to "see" Jesus with greater historical clarity. No. As important as that "seeing" is, it's only the first liberating step. Jesus was a spiritual revolutionary whose "program" was to transform human consciousness and all human institutions. The "kingdom" he invited us to enter is a deeply and pervasively revolutionary realm. Yet, even though the twenty-first century imagines itself a world free of peasants, it's nearly impossible to grasp the kingdom of God, to understand its depth and dynamics, without some grasp and understanding of the peasantry or of the extractive relationship between civilized elites and peasants. Although the work of Crossan and Borg doesn't take us back to the origins of civilization, of how a bandit aristocracy impounded the agrarian village, their engaged and sympathetic understanding of the connection between the peasantry and Jesus' kingdom of God opens the door to a deeper interpretation. A spiritual assimilation of the profoundly important connection between the kingdom and the liberation of peasant life prepares the ground for a revolutionary understanding of civilization and the impoundment of agrarian culture.

The kingdom of God is not a concept in a vacuum, nor is it a term without a context. It's not a vaguely "spiritual" ideal floating in social ungroundedness or levitating on a cloud of wishful thinking. It has to be seen in relation to the actual historic groundedness of peasant life and peasant culture, and it has to be integrated into prehistorical scholarship—excuse me, the scholarship of prehistory—and the origins of civilization.

And yet it's more than just the peasant understanding of civilization or our understanding of the peasantry in relation to civilization that's at issue here. The kingdom of God transforms the best and finest and most communally supportive practices of oppressed peasant life and imbues them with divine radiance; it lifts them into perpetual Jubilee, and it proposes to do this *now*. To see civilization from *this* perspective is to recognize the congealed hierarchical system of oppression for what it is, whose main and sustained features are wealth extraction and wealth concentration, coerced labor, and legal violence—all in the service of elite privilege. Civilization uses its monopoly on violence to achieve and sustain its monopoly on wealth and power. It forces into being an economic pyramid, though remarks from both Marcus Borg and Dom Crossan show how poor and inadequate an image "pyramid" really is.

To be lured by and drawn to the civilized standard of living—even as it hides behind the fig leaf of "democratization"—is to be drawn away from the ripe but existentially difficult kingdom of God. While, in one respect, the kingdom of God depends and relies on sustained spiritual renunciation, civilization supreme is self-advancement and self-promotion supreme, even though it can promote "sacrifice" in moments of need or crisis. The destruction of the peasantry opens the door to a totally civilized economy, that is, to a globalized economy of technological rationalization and total market penetration. Its intent is to eradicate sufficiency, sharing, stewardship and servanthood (except for "volunteer" and "sacrifice" condescension) and suspend us all in its organizational webwork. The kingdom threshold can be crossed only with childlike simplicity, with a depth of trust in the omnipresent, unlimited compassion of Spirit, whose abstract "existence" we may otherwise aggressively assert, insist we "believe in," demand "obedience" to, but, in the functional conduct of our normative alignment with "civilized values," totally and absolutely disbelieve and scorn.

Behind the glittering civilized idolatry of our smug "belief" lies the global genocide of Creation's peasantry. And that raises the exquisitely painful question of what's meant by resurrection.

In one of his essays ("The Passion, Crucifixion and Resurrection") in *The Search for Jesus*, Dom Crossan says that, in the thought of the Apostle Paul, the resurrection of Jesus starts the "general resurrection." For Paul, "the resurrection of the dead has begun" and "the end of the world *had already begun*...."[26]

I suspect it is foolish and foolhardy in the extreme to look for or demand empirical definition for this terminology. I further suspect its relevance can be grasped and assimilated only as one has crossed the threshold of the kingdom and tasted both its fruits and its sufferings. But by keeping religion in a neat little ecclesiastical box decorated with comforting streamers of "belief," we evade and avoid struggling with the existential meaning (or madness) of such resurrection language. Was Paul a madman irresistibly drawn along on the maniacal afterglow of an even madder madman? That's one option—an option we can gratefully, if also uneasily, choose as we side with a putative and unlimited "humanizing" potential of evolving civilization and its endless technological "progress."

The other option—voting for Jesus—is as existentially foolish now as it was nearly two thousand years ago: just as foolish and just as unimaginably real. A vote for Jesus is a vote for servanthood and stewardship, sufficiency and sharing, a vote for the resurrection of the peasantry and for the unfolding of the kingdom of God.

In another of his *Search for Jesus* essays ("The Infancy and Youth of the Messiah"), Dom Crossan quotes the second-century Roman Celsus' attack on Christianity:

> "This savior, I shall attempt to show, deceived many and caused them to accept a form of belief harmful to the well-being of mankind. Taking its root in the lower classes, the religion continues to spread among the vulgar: Nay one can even say it spread because of its vulgarity and the illiteracy of its adherents. And while there are a few moderate, reasonable and intelligent

people who are inclined to interpret its beliefs allegorically, yet it thrives in its purer form among the ignorant."[27]

And then Dom Crossan comes to his own conclusion:

> In summary, then, it is not enough to keep saying that Jesus was not born of a virgin, was not born of David's lineage, was not born in Bethlehem, that there were no stables, no shepherds, no star, no Magi, no massacre of the infants, no flight into Egypt. All of that is absolutely true. But it still begs the real question, which is, then as now, where *you* find the divine manifest on this earth. Is it in Caesar, or is it in Jesus? Is it in imperial grandeur or peasant poverty? Is it in domination and subjugation of others from the top down, or is it in the empowerment and liberation of others from the bottom up? That's the real question.[28]

Whom will we vote for with our lives—Caesar or Jesus? Or do we prefer to skip the election and just go shopping?

Let's add one more thought. For all the down-to-Earth, gritty recognition of a "brokerless kingdom" in Dom Crossan's scholarship, there is one particular word usage that is totally unexplored and utterly conventional. That word is "pagan." I will give merely one very unremarkable example. In the Epilogue to *The Historical Jesus*, Dom Crossan refers to "the fictional *Letter of Aristeas to Philocrates*" whose "Jewish author, speaking fictionally through a pagan, says that the Jews 'worship the same God—the Lord and Creator of the Universe, as all other men.' . . . That quite extraordinary admission from a Jew about paganism is matched in the next century by a pagan, Marcus Terentius Varro, speaking about Judaism."[29]

"Pagan" in this familiar usage is both quasi-prejudicial and pseudo-objective. That is, it includes all the religions, social systems, and persons who were not Jewish, Christian, or part of what we now call the Judeo-Christian tradition. In, perhaps, the most typical scholarly usage, "pagan" meant the Romans, the Roman Empire, the Greco-Roman tradition, and so on: "pagan civilization."[IV] "Pagan" in its Latin root means country district or country dweller—a root it shares with "peasant." So the rise of

---

IV. In his *Everything Must Change*, on page 137, Brian McLaren more or less repeats this convention: "When we hear 'pagans,' we typically think of the irreligious in contrast to the religious, but the Greek word *ethne* (the original word translated 'pagans') simply means Gentiles, or non-Jews, who in Jesus' day are none other than members of the Roman Empire." Well, *ethne*, ethnic, and ethnos appear more related to *ethical* than to pagan, although one can see how "ethnic" in particular could be construed as "pagan." Nevertheless, "ethnic" is simply descriptive, while "pagan" comes loaded with demonic energy—less a value-neutral description than a vile epithet.

"pagan" from rustic origins to empire status is therefore a kind of semantic rags-to-riches story, without even an apparent bandit chapter.

But how did a term meaning a country person or country place gradually come to mean, in conventional Christian terminology, a wicked unbeliever (or a believer in false religion), and, in scholarly usage, a *civilized* system (or some portion of it) with non-Jewish or non-Christian origins? How, in other words, does a term that, in its root, is synonymous with peasant life or rural place come to stand for or in any way be meaningfully descriptive of emperor or empire? How in the world did this happen?

Here, following Professor Borg's lead, I propose four dynamics—peasant, civilization, pagan, Christianity—all, finally, in relation to the kingdom of God. If we were to rank these terms in the moral order of conventional understanding, pagan would be at the very bottom. It would be unconditionally the least moral or godly. At least that would be true in Christian teaching and doctrine, where "pagan" represents false belief and false conviction, a falsity probably grounded in "nature worship" or even "devil worship." Its spiritual connotation is, finally, evil.

Somewhere along the line "pagan" became a term of grudging admiration for "pre-Christian" civilization: "pagan civilization." It became a scholarly and intellectual convention. The magic was in the capacity of the term to enable its users to simultaneously admire these early civilizations while shielding themselves from spiritual contamination. "Pagan" worked like a doctor's mask and rubber gloves. One could perform delicate scholarly surgery on the past without serious risk of infection. Over against a censorious Christian establishment the term implied a comforting protectiveness: "See, we are examining the past, but we recognize its spiritual wickedness and we are taking precautions." Those civilizations were bad because they were "pagan," not because they were civilizations.

The essentially unspoken and unarticulated element within "pagan civilization" is the implied assertion that ours is a *Christian* civilization. Whatever may have been in err with *pagan* civilization has now been Christianized and is therefore right with God.

If "pagan" in its basic meaning is at the bottom of the semantic barrel, "peasant" is one step up. A peasant may be a plodding sort of person, unsophisticated and a dullard, but he is not necessarily evil or wicked. If "pagan" is wicked, "peasant" is merely stupid. He has cow crap on his shoes, mud between her toes.

Now the next step—a huge step—in this ranking is more difficult. Is "civilization" above or below "Christianity"? The fair and honest answer is that we are all familiar with a term like "Christian civilization" but almost totally unfamiliar with "civilized Christianity." That is, when push comes to shove, "Christian" comes out in dominant convention as a mere adjective, a sign, a pointer, an indicator of something greater and more substantial. In the push for the top spot, "Christianity" comes in second. "Civilization" gets to sit on the semantic throne. At the very most we are permitted by convention to say that civilization has been "Christianized."

I will quote here the *entire* entry for "pagan" from my twenty-four volume set of the *Encyclopaedia Britannica*:

> *PAGAN*, a heathen, one who worships a false god or false gods, or one who belongs to a race or nation which practices idolatrous rites and professes polytheism. In the early Christian Church *paganus* was applied to those who refused to believe in the one true God. It thus, of course, excluded Jews. In the middle ages, at the time of the crusades and later, "pagan" and "paynim" (O.Fr. *paenime*, Late Lat. *paganismus*, heathenism or heathen lands) were particularly applied to Mohammedans, and sometimes to Jews. It was in the rural districts that the old faiths lingered, and thus it is assumed that the Latin *paganus* (villager) arose after the establishment of Christianity, but Tertullian (c. 202, *De Corona Militis*, xi.), has a sentence suggesting that the "soldiers of Christ" dubbed the non-Christians *pagani*, referring to the raw, half-armed rustics.[30]

That's it. That's the entire entry. One would think the peculiar historic evolution of "pagan" would have elicited a hugely longer entry in the insatiably curious *Encyclopaedia Britannica*. The word, after all, simply squirms with implication and innuendo. It represents, among other things, the lascivious underbelly of natural desire that Christianity struggled so ardently to stuff securely into a godly chastity belt. That the entry is so brief, so painfully succinct, so self-effacing and unaccountably modest, suggests a certain (but rather intense) intellectual discomfort and scholarly embarrassment handled discreetly by omission and avoidance, rather like a ripping good fart at an elegant dinner—obvious enough to have made the news, but only in the gossip column.

Now, what about the kingdom of God? Dom Crossan and Marcus Borg have already underscored Jesus' linkage and identification with the peasantry. (Crossan has told us of the "strategy for building or rebuilding peasant community on radically different principles.") The kingdom of God (whatever else it may be) is the spiritual and cultural transformation of peasant consciousness and peasant life into an ongoing Jubilee of sharing, healing, commensality, equality, and just plain joyful living: the kingdom of God means an earthy, spiritual, wholesome *life without civilization*, and the living revolution is to start *now*.

Here is Dom Crossan from his Epilogue in *The Historical Jesus*, from a brief section subtitled "Jesus and Christianity":

> It is hard, indeed, not to get very, very nervous in reading this description of the imperial banquet celebrating the Council of Nicaea's conclusion:
>
> "Detachments of the body-guard and troops surrounded the entrance of the palace with drawn swords, and through the midst of them the men of God proceeded without fear into the innermost of the Imperial apartments, in which some were the Emperor's companions at table, while others reclined on couches arranged on either side. One might have thought that a picture of Christ's kingdom was thus shadowed forth, and a dream rather than reality." (Eusebius, *Vita Constantini* 3–15; Brown 1982: 16)
>
> The meal and the Kingdom still come together, but now the participants are the male bishops, and they recline, with the Emperor himself, to be served by others. Maybe, Christianity is an inevitable and absolutely necessary "betrayal" of Jesus, else it might all have died among the hills of Lower Galilee. But did that "betrayal" have to happen so swiftly, succeed so fully, and be enjoyed so thoroughly? Might not a more even dialectic have been maintained between Jesus and Christ in Jesus Christ?[31]

Here, in this (also agonizingly brief, succinct, self-effacing, modest— and wary?) grieving over Christianity's capitulation to Empire, we have at least an implicit answer to the question of the relationship between Christianity and the kingdom of God. Dom Crossan has given us more than a hint: Christianity without the kingdom of God is Christ devoid of Jesus.

So, if the kingdom of God is fully at home among peasants, is the enemy weed and stinky leaven that civilization hates, is the denied and neglected "illegitimate" ancestor of a conventional Christianity trying desperately to be metaphysically respectable, what is the relationship between the kingdom of God and "pagan"? The answer to this question is of necessity less clear precisely because "pagan" is such an ancient scapegoat term. "Pagan" is, historically speaking, a stocky and burly character with three heads. In the center, the oldest face looks remarkably like that of a common peasant, sturdy, reliable, and strong. On the left, it's the face of a demon or a devil, very sinister. On the right, it's a "noble" amalgam of

Alexander and Caesar, Socrates and Cicero, wily, sophisticated, and wry. If we nip off the right and left heads—the right because it's a projection of idealized civility, and the left because it's a scapegoat projection of wickedness—we are left with peasant's twin. So we can say as a broad generality that as peasant correlates to simple, subsistence agriculture, pagan correlates to healthy, natural ecology. We can go on to say that if servanthood and stewardship are in fact the earthly, ethical legs of the kingdom of God, then servanthood correlates to the rebuilding of peasant community while stewardship correlates to the healing of Earth's ecology. And this, too, is why the kingdom of God is Green, and why the kingdom of God implies the resurrection of both peasant and pagan—a resurrection that's also a transformation.

Perhaps another way to frame this question is like this: What is the *theology* of the kingdom of God? If Jesus' "Abba" is a Spirit of infinite grace, mercy, and compassion, a bottomless pool of forgiveness and redemption, then the kingdom of God is not a "state" of harsh ideological surveillance where a pledge of allegiance (or a urine drug test) is a requisite part of creedal conformity. The ethics of the kingdom are rooted in the regular return to being in the presence of Spirit: to *be* in that presence, and to allow *being* in that Being to transform and cleanse one's spirit. The result of this cleansing are persons who are deeply reverential toward Spirit, eager and willing to live simply in a context of communal sharing, hugely respectful toward the Creation in which Being has manifested its beings.

Here are people who are happy, who are largely content, who love the smell of wood smoke and the taste of an unpoisoned apple, who are so deeply grounded in their relationship with Being that they are not prone to moralistic judgmentalness in regard to labels. If there is a word that evokes suspicion, it is the old one that the wolf used for himself, when he masqueraded as a lamb, back in the Old Days when animal farm really meant its opposite—civilization.

Notes

1. Borg, "Palestinian," 44–45.
2. Borg, "Palestinian," 48–49.
3. Crossan, *Historical*, xxvii.
4. Crossan, *Historical*, xxxiv.
5. Crossan, *Historical*, 42.

6. Crossan, *Historical*, 121.
7. Crossan, *Historical*, 169.
8. Crossan, *Historical*, 171.
9. Crossan, *Historical*, 172.
10. Crossan, *Historical*, 174.
11. Crossan, *Historical*, 193.
12. Crossan, *Historical*, 220–21.
13. Crossan, *Historical*, 263.
14. Crossan, *Historical*, 266.
15. Crossan, *Historical*, 275.
16. Crossan, *Historical*, 278–79.
17. Crossan, *Historical*, 280–81.
18. Crossan, *Historical*, 292.
19. Crossan, *Historical*, 298
20. Crossan, *Historical*, 302.
21. Crossan, *Historical*, 341.
22. Crossan, *Historical*, 344.
23. Crossan, *Historical*, 349.
24. Crossan, *Historical*, 404.
25. Crossan, *Historical*, 422.
26. Crossan, "Passion," 122.
27. Crossan, "Infancy," 75.
28. Crossan, "Infancy," 75–76.
29. Crossan, *Historical*, 419.
30. *Encyclopaedia Britannica*, Vol. 17, 26.
31. Crossan, *Historical*, 424.

15

The Gathering Globalization of Disaster

It seems that the very basic "kingdom of God" that Jesus taught, practiced, and enacted went through a number of phases in subsequent history; it also seems these "phases" are discernible. I don't pretend for a moment the following list is conclusive or complete—but here goes.

First, looking through the window of the Gospels, one sees stories that tell, largely, of peasant gatherings based on intense spiritual hunger and personal commitment, on voluntary poverty (in the midst of pervasive involuntary poverty), on incidences of bewildering healing (both physical and mental), on shared meals and banquets, on stories told to crowds, on angry and pointed denunciations of the rich and comfortable, of leaders, rulers, professors, lawyers, and priests. This is a magnetic social movement, a revolution of charismatic energy, that invites all those, especially the poor and outcast, who have little to lose but their misery, to a communal movement based on intensive sharing, compassion, healing, and love. This is cultural revolution, character transformation, and a certain sort of spiritual redemption all rolled into one fantastic bundle of creative energy. That's phase one.

Second, the religious and civil authorities conspire to kill Jesus, this extraordinary charismatic person of incredible bravery and unflinching honesty, and the remnants of the community he created are thrown into confusion, onto their own inadequate resources. They are scared and they hide, but they hang on in ways that perpetuate some degree of kingdom awareness, and they strive to realize this awareness in local groupings of ongoing community.[1] They tell stories about this incredible, charismatic Jesus, and these stories morph and spread.

Third, as the teaching of and about Jesus expands, primarily throughout and along the perimeter of the Mediterranean basin ("a dramatic thrust along major trade routes to urban centers hundreds of miles apart," as John Dominic Crossan has put it), an increasingly abstract and metaphysically elusive view of Jesus tilts toward a salvation figure whom the local community gathers to worship—with, no doubt, varying levels of functional daily community intensely needed and actually practiced, especially under Roman persecution, as real community remained vital for survival. Jesus' extraordinary existential groundedness begins to get idealized as otherworldly salvationism.

Fourth, by the first quarter of the fourth century, the "orthodox" church (i.e., the church that honed its particular set of doctrines and creeds in ways that achieved dominance) has climbed into bed with empire, and the cross becomes an emblem of adoration and power. Augustine soon emerges to instruct the faithful on the God-given nature of empire, on the role of angels of light and angels of darkness, and on the "two cities"—the City of God and the City of Man. Augustine pumps Jesus full of Greek philosophy. Within a few centuries, the church is not only the inheritor of a collapsed and overrun empire, it is also the owner of huge estates and vast tracts of land. The monastic movement grows and strengthens to some degree in protest against the political empiring of the church.

Fifth, the monastic movement itself becomes both the carrier of the kingdom of God as community and a formalized substitute for it. That is, the "kingdom of God" retreats to a self-selecting order of chaste monastics strictly organized and doctrinally vetted who no longer function as communal leaven among the peasantry with sex, babies, and families. Church fathers have apparently beaten erotic energy out of the kingdom of God. The church becomes a highly complex institutional structure with a huge amount of political power controlled essentially by male aristocrats. The church has also become misogynistic and theocratic. It most decidedly does *not* seek the kingdom of God on Earth, at least if one means by "kingdom" the committed and sustained practice of radical servanthood and radical stewardship in a yeasty folk context. That is, just as Jesus was transformed from a cultural and spiritual revolutionary to a divine keyhole through which to get a glimpse of heaven, so the "church" was similarly

---

1. See especially Ched Myers' *Binding the Strong Man: A Political Reading of Mark's Story of Jesus* for a very down-to-Earth reading of what the kingdom of God is about.

transformed from oppressed peasants yearning for wholesome liberation to a waiting room crowded with expectant otherworldly tourists.

Sixth, the reformation of the sixteenth century hammered huge areas of the Roman church, closed and expropriated monasteries, worked (perhaps inadvertently) to the maximum advantage of a strongly emerging merchant class, and—with the exception of various Anabaptist sects—never remembered or promoted the meaning of the kingdom of God on Earth as it focused almost totally on successful inclusion of church members in heaven after death. Right doctrine—believing the right things in the right way—became uppermost for the attainment of salvation. The Bible—*sola scriptura*—told us all we needed to know.

Seventh, with the ancient imperialism of civilization metaphysically reenergized by the religious imperialism of Christianity, and with the technical means at hand by which to extend and enlarge these imperialisms, the entire "pagan" world was aggressively explored and incrementally conquered. Technical instrumentalities progressed by leaps and bounds. Virtually all the world's people were influenced if not overtly dominated by Western ideas, political systems, and new technologies.

Eighth, a new, massively magnified affluence in the West also served to help secularize intellectual thought, compounded by scientific discoveries (as in astronomy) that served to discredit and subvert religious certainty. Secular thought began to shake itself free from the controlling constraints of religious ideology and even on occasion denounced religion as a mass hallucination that a) was clung to out of an unwillingness or inability to face the facts of life and death or b) was used as a controlling bogeyman by the ruling class to keep the lower classes under control or c) was employed by arrogant males to keep females in subjection or d) all of the above. Christianity had become overwhelmingly a religion wrapped in a controlling myth.

Ninth, for some revolutionaries who despised the self-serving manipulations of the ruling class, the objective was to form an armed revolutionary party and seize the state (that is, civilization's control center) and impose a regime of equality and sharing. Where this determined impulse for sharing and equality comes from was, as a rule, not much discussed because such a discussion opens the door to the spiritual roots of ethics, and spirituality has been unconditionally rejected because of its association with state church rigidity and its cloying otherworldliness. Unable to talk about the kingdom of God as a political, economic, and sociological concept and program—not only because it always contains that rationally uncontrollable spiritual element, but also because of its sustained atrophy within an ecclesiastical Christianity now held in contempt—revolutionary materialists typically fell back on two sharp-edged and dangerous words. Those words were "utopian" and "civilized." The revolutionary future was to be both civilized and utopian. And this, in turn, means that civilized consciousness had become so demographically infiltrating that even the most determined of revolutionaries—especially the most determined revolutionaries—failed to recognize that their utopian projections really needed to be talked over with peasants whose eutopian liberation should unfold from its generations-long suppression by aristocracy. Utopia sent eutopia to Siberia, instead. And, without a peasant soul, the Communist Party of the Soviet Union died of utopian asphyxiation. (If only Gorbachev could've had more time!) But Soviet utopianism switched to mafia fascism—a logical step in bitter disillusionment, given its absence of peasant depth—just as Clinton and Gore sent in the "shock doctrine" Chicago Boys to build a new political-economic behavioral model while Yeltsin got drunk and Putin waited in the wings. The problem with Marxism and communism was and is their utopian superegos. They needed a whole lot more eutopian id. But, like all good altar boys, communist revolutionaries learned to have no truck with pagans.[II]

Tenth, "progress" and "development" as utopian terms means that Christianity as a whole, as a generality, has also fallen prey to civilized utopianism. Christianity no longer knew how to be uncivilized or think in a noncivilized manner. Utopia is an ideal to be achieved whereas the kingdom of God is a reality to be lived. Having an ideal to be achieved enables one, with the finest rationalizations possible, to mercilessly betray the immediate reality. Utopia is the building of a (heavenly) system that, once completed, will permit us to share, to love, to be happy, and to be fully human. But first we have to build that (heavenly) system, and that involves deference, coercion, denial, repression, oppression, exploitation, and sacrifice. Utopia is secularized but postponed transcendence.

Jesus, I believe, never left the present moment. Never. If we dare to call him God or Son of God, it's because he so fully embodied Living Presence. So, as all the deferred existentiality of all utopian civility congeals in our time as demonic sublimations—what Altizer has called "the consummation of all oppressive power"—the gathering globalization of disaster, with its horrific contradictions, oozes violence and destruction.[III]

II. See Chapter 14, "The Conscious Id," in my *Green Politics Is Eutopian*.

III. See Norman O. Brown's *Life Against Death: The Psychoanalytical Meaning of*

The Kingdom of God Is Green

Meanwhile, the kingdom of God, with its maddeningly tranquil presence, is waiting to be discovered and enacted among us.

And that is to say—I address this to ethical socialists above all—socialism in its finest sense is only possible with the conscious, reverential, and humbling inclusion of a spirituality that lies beyond rational dissection and intellectual control, an inclusion that willingly recognizes and acknowledges Spirit as the ground of our lives and ethical consciousness. I am not, strictly speaking, selling Christianity here, thrusting a tract into your hands, or demanding to know whether you have Jesus in your heart. (Though Christianity is, by far, my dominant religious influence, with a dash each of Taoism and Buddhism and some almost too-elusive-to-meter impacts from Native American spirituality.) I am saying that Spirit *is*, and our capacity to achieve elegant human community—both servanthood and stewardship—cannot be realized without our deliberate and voluntary humbling before Mystery. This is the eye of every needle. We ourselves are the camels who must be coaxed through. Denial of Spirit means, ultimately, denial of the revolution that is also the "kingdom of God."

Our pride, self-esteem, and self-regard (not to speak of our massively corrupted political self-inflation, with its climate-changing economy and End Times weaponry) need to be dropped, let go of, renounced, and abandoned. This is both the personal and political meaning of repentance.[IV] Utopian identity (including its secular side of reflexive atheism) is an addiction that must get scraped away—perhaps painfully—as we consent to be spiritually needled. Without this wakeful transformation we only contribute to and line up for necropolis, the city of the dead, the ultimate utopia of violence, destruction, and disaster.

History, especially Part Five ("The Excremental Vision," "The Protestant Era," and "Filthy Lucre").

IV. Although expressed a little too piously for my taste, the Quaker Thomas R. Kelly, in *A Testament of Devotion*, page 96, asks if we want to live a life that "is transformed and transfigured and transmuted into peace and power and glory and miracle? If you do, then you can. But if you say you haven't the time to go down into the recreating silences, I can only say to you, 'Then you don't *really* want to, you don't yet love God above all else in the world, with all your heart and soul and mind and strength.'" No doubt, for us, it's not so much a lack of time as lack of inclination, although we very much need to bear in mind that ecclesiastical condescension toward the kingdom of God provides a comforting nest in which our collective disinclination can safely sleep and has securely slept.

16

Packed in a Comforting Mythology

EXTRA! IS THE MAGAZINE of Fairness and Accuracy in Reporting (FAIR), a "media watch" group, and, in its April 2002 issue, there is a long piece by Noam Chomsky, probably the most widely known living "radical" in the United States. Chomsky's article, entitled "The Journalist from Mars," was really a talk he presented at FAIR's fifteenth anniversary celebration, in January of 2002, in New York City.

The core of Chomsky's talk has to do with the definition of "terrorism," and he quotes from official U.S. documents: "terrorism" is the "calculated use of violence or the threat of violence to attain goals that are political, religious or ideological in nature . . . through intimidation, coercion or instilling fear."[1]

Chomsky's point is that this definition, though quite clear, invariably makes trouble, by virtue of its inclusiveness, for any government, including our own, and that it therefore becomes necessary to redefine "terrorism" as "the terrorism that *they* carry out against *us*, whoever we happen to be. As far as I know, that's universal—in journalism, in scholarship, and also I think it's a historical universal; at least, I've never found any country that doesn't follow this practice."[2]

I accept and concur with this analysis and have nothing more to add. I have only one point to make, and that indirectly concerns the term "radical."

Chomsky is considered, by those who engage in political definitions, to be on the "extreme" Left. (He's too far Left for *The Nation*, for instance, which tended, prior to the 2000 elections, to hammer Nader and promote Gore, and, in the aftermath of September 11, to support and even urge on the "war against terrorism.") I believe Chomsky is a socialist, but it's not his socialism that makes him "extreme" in the eyes of the conventional Left.

The Kingdom of God Is Green

What makes Chomsky "extreme" is implied in his citing the definition of "terrorism." In the definition provided by our own government, terrorism is the calculated use of violence or threat of violence to attain political, religious, or ideological goals. By this definition, terrorism is the real working ingredient in Arnold Toynbee's War and Class, the driving fear or threat of violence that makes the system run. Terror is, when all is said and done, the controlling electrical current of our civilization and all previous civilizations—what Noam Chomsky calls a "historical universal" and what Arnold Toynbee seems inclined to call a "disease."

Chomsky's "extremist" nature boils down to his penchant for moral and ethical consistency, and it's this unflinching consistency that causes conventional Leftists to get inflamed—not to mention Centrists and "conservative" Rightists. Chomsky asks us to imagine an intelligent Martian who is a journalist on Earth, reporting back to Mars. He's bright, he's honest, he's unfamiliar with Earth politics, and he wants to get it right in his reportage. Here's Chomsky from his talk:

> A good Martian reporter would also want to clarify a couple of basic ideas. First of all, he'd like to know what exactly is terrorism. And, secondly, what's the proper response to it. Well, whatever the answer to the second question is, that proper response must satisfy some moral truisms, and the Martian can easily discover what these truisms are, at least as understood by the leaders of the self-declared war on terrorism, because they tell us, they tell us constantly, that they are very pious Christians, who therefore revere the Gospels, and have certainly memorized the definition of "hypocrite" given prominently in the Gospels, namely the hypocrites are those who apply to others the standards that they refuse to accept for themselves.
>
> So the Martian understands, then, that in order to rise to the absolutely minimal moral level we have to agree, in fact insist, that if some act is right for us then its right for others, and if it's wrong when others do it then it's wrong when we do it. Now that's the most elemental of moral truisms, and once the Martian realizes that, he can pack up his bags and go back to Mars. Because his research task is over. He would be unlikely to find a phrase, a single phrase in the vast coverage and commentary about the war on terrorism that even begins to approach this minimal standard. Don't take my word for it; try the experiment. I don't want to exaggerate—you can probably find the phrase now and then, way out at the margins, though very rarely.

> Nevertheless, this moral truism is recognized within the mainstream. It's understood to be an extremely dangerous heresy, and therefore it's necessary to erect impregnable barriers against it, even before anybody exhibits it, even though it's so rare. In fact, there's even a technical vocabulary available in case anybody would dare to engage in the heresy, to involve themselves in the heresy that we should abide by moral truisms that we pretend to revere. The offenders are guilty of something called moral relativism—that means the suggestion that we apply to ourselves the standards we apply to others.[3]

Noam Chomsky, the extreme radical, is clearly guilty of the "most elementary of moral truisms." He's a dangerous heretic. That is, he's guilty of telling the truth, of believing in the humanity of peace and sharing and nonviolence, and of believing in it as much for "us" as for "them." He's not into giant windbag terms like "civilized values" or metaphysical monstrosities like the "clash of civilizations." He has no truck with the "us" against "them" of "civilization versus terrorism." He looks at what I suppose we are obliged to call "human nature" and concludes we are safer taking our chances with sharing and stewardship than continuing on with the underlying greed and power games of Class and War. In that respect, Noam Chomsky is a true conservative. He wishes to *conserve* Earth's ecology and a humane culture of reverence and sharing.

His message is really quite simple. If we're serious about wanting a humanity more at peace, we who have (by far) the most weapons have to disarm and become more peaceful; we who have the most wealth to share have to share it; we who consume the most must restrain our consumption; we who pollute the most must reduce our pollution. These are clear and obvious examples of the moral truisms we say we believe in—it's all there, explicitly and implicitly, in the Gospels—do onto others as you would have them do onto you—and that we ignore, deny, and try to weasel out of at every inconvenient moral dilemma.

The civilized Left hates Chomsky because he has, at least implicitly, stepped beyond the hypocritical conventions of civilization and has left both materialism and idealism behind. Noam Chomsky, it seems, is too spiritually radical for the "radicals." He is certainly too ecologically conservative for the "conservatives."

There's no great mystery here—except for the mystery of why we cling to Class and War with such intensity. Perhaps we are more afraid of what the world would be like without civilization than we are of what civilization is most certainly going to do to the Earth and to us if we keep clinging to it as our metaphysical savior. We seem to *like* the distance separating us from the kingdom of God. We may desperately desire the continuance of our egos in a life after death, but it's crucial to keep our eyes on the prize and not get waylaid by any "kingdom of God" crap about servanthood and stewardship on Earth. We are comfortably packed in a mythology that assures us by whispering, over and over, how civilization, while righteously punishing those who resist its civility, humanizes backwardness and spiritualizes primitivity. Christianity just skims cream off the top. Meanwhile, Class and War ratchet up their power and break down all resistance to the spread of their infection.

Toynbee was close to having it right. Class and War are not the diseases of Civilization. Civilization is the disease produced by Class and War. This is a disease to which (if I can bend the metaphor) we've become addicted. We drug ourselves with hypocrisy. On Mars they see this quite clearly.

II

Rereading this essay in July, 2003, I find that the sentence "Perhaps we are more afraid of what the world would be like without civilization than we are of what civilization is most certainly going to do to the Earth and to us" jumps out at me. This jumping occurs, in part, because a friend who was a World War II infantry paratrooper dropped at Normandy, recently loaned me a book by David Kenyon Webster, also a Normandy paratrooper, a book entitled *Parachute Infantry*. It's a very straightforward narrative of war as seen through Webster's eyes.

The point here is an anecdote told by David Webster. His lieutenant orders him and a few other men to position themselves on a mound that is, in their conviction, about to be blown to bits by American artillery. Webster protests. Lieutenant Peacock orders. The men obey. Here's the paragraph:

> We are disciplined, so we lie here and take it, because, in the end, we are more afraid of defying the authority of an officer, backed up by the whole Army and a court-martial composed of officers like him, than we are of death by shell fire. Discipline is fear, not leadership, and we are afraid—not of Peacock but of the irresistible force that he represents. Afraid for our lives, we are more afraid of the system that holds us in thrall, and so we lie here and wait to be killed, because an officer tells us to lie here.[4]

This is one extremity of "the system that holds us in thrall"—lying in wait of death rather than walking away. I realize Webster refers here explicitly to the military (an aristocratic fist that has never left us); but the "irresistible force" is even more fully lodged in civilization itself. The officer, after all, is a military aristocrat. The infantryman is a mere peasant. The army is the fist of civility. Discipline is fear. We live out our obedience and acquiescence at a level mostly below consciousness, as we presume the reality we experience—even when manifestly unjust and unmistakably irrational—is God's will or a law of nature. That our social and political reality is overwhelmingly a convention of such fiction is not only beyond our ken; but, when faced with its lethality, we will obey out of fear and habit. We will lie here and take it.

Conversely, Robert Caro relates an anecdote about Mississippi segregationist Senator Theodore Bilbo who, in a 1947 self-published book entitled *Take Your Choice: Separation or Mongrelization*, said it is "better to see civilization 'blotted out with the atomic bomb' . . . than to see it slowly destroyed in the maelstrom of miscegenation, interbreeding, intermarriage, and mongrelization."[5]

In other words, soldiers—mere peasants, a specially trained kind of slave—may die by the thousands and even millions as pawns in the game of civilized warfare, but it is better that civilization (and, presumably, the entire inhabitable Earth) be utterly destroyed by atomic weapons than that the infinitely superior white race be "mongrelized" by intermarriage with inferior breeds. This is beyond insult, perhaps even beyond criminality. It's psychotic with supremacy and superiority, possessed by a kind of "holiness" that's a vicious, cowardly, superego mask for a depth of desire denied to the point of ideological racist doctrine.

What we see here are very real snapshots of what it means to be "in thrall" to a perfectionist system of ideological and religious imaginings with explicit supremist convictions. Their dreadful relevance is epitomized in every atomic weapon, in every denial of climate change, in every triumphant utterance in behalf of the "only true religion," in every political lie that perpetuates "civilized values" or Manifest Destiny.

On Mars they see this quite clearly.

Notes

1. Chomsky, "Journalist," 12.
2. Chomsky, "Journalist," 12.
3. Chomsky, "Journalist," 11–12.
4. Webster, *Parachute*, 145.
5. Caro, *Master*, 690.

17

The Superlative Proportions of Our Self-Inflation

WE SEEM TO LIVE, largely, in a kind of Norman Rockwell atmosphere—neither quite real nor yet a cartoon. Perhaps the word I'm looking for is not "caricature" but "mythological." We see the world through the tinted glasses of mythology. Probably this is to some extent unavoidable.[1] So much of our consciousness is shaped and influenced by cultural forces (in which mythology is fully and inextricably threaded) that the question is not nearly so much the purging of mythology from consciousness as it is the creative and compassionate wrestling with mythology in a sustained atmosphere of spiritual humility and intellectual courage. We need to bring the discrete and selectively hidden aspects of myth into the radiant transparency of truth.

Former House Minority Leader Dick Gephardt of Missouri expressed a certain mythology in the Democratic Response following President George W. Bush's State of the Union address in January, 2002. Gephardt

---

1. Perhaps we can remember that Gil Bailie, on page 33 of his stimulating *Violence Unveiled*, says "the root of the Greek word for myth, *muthos*, is *mu*, which means 'to close' or 'keep secret.' *Muo* means to close one's eyes or mouth, to mute the voice, or to remain mute. Myth remembers discretely and selectively." Bailie goes on to say that "the term 'muse' is derived from the same root as the word 'myth' [and that the] Muses make culture possible by providing it with its myth—an enchanting story of its founding violence." This is brilliant commentary, but I don't believe all myth originates from an enchanting story of founding violence—that's way too Manichean an assertion for me—but how else are we to understand the fierce exclusion of the feminine in Abrahamic divinity except as a consequence of the aristocratic/civilized overthrow of the agrarian village, an overthrow that fundamentalism denies ever happened and that civilizational mythology says was a necessity for human progress?

The Superlative Proportions of Our Self-Inflation

said—I heard this on the radio, and I think it's an exact quote—"America is the greatest country that has ever existed in the history of the world."

Such a statement is, of course, a kind of shameless mythological pandering to a political audience presumably expecting to have its superiority stroked, as if it were empirically true and historically verifiable that "we" are the "greatest" country—and therefore the greatest people—not only in the world but in the *history* of the world. (Modesty, apparently, is not a virtue worth cultivating.)

There is always the problem of whom it is Gephardt (and other pander bearers like him) truly is addressing. The audience is, of course, a big one—the "public." But there is another problem here, and the doorway into this problem is in a building called "Discussion and Debate." The doorway is, if not in every instance always locked, certainly always guarded—carefully. The question then is: Who are the guards? Who determines what's permissible in public discussion and debate? Who protects myth?

The guards wear distinct emblems. The most powerful emblem is the dollar sign, the icon of wealth and of the power wealth brings. The second emblem is the flag, behind which is the camouflage "team" of warriors whose capacity for violence insures the superiority of the dollar and the supremacy of the flag. The third emblem, evoked only in its narrowest Augustinian empire mythology sense, is the cross—asserting the powerful idea that God and Jesus have done it all, and all we need do is be passive-aggressive believers. The dollar, the flag, and the cross constitute the trinity of America's secular religion.

A further question puzzles me: What are the common roots of this mythology? How is it that a national politician not only *can* say such an insane and asinine rubber-ducky thing, but apparently feels he *has* to? What conventions in us was Gephardt aiming at?

Well, he obviously expected the majority of us would agree with him. So he was saying something, presumably, that would make us feel positive toward him as a political "leader." He was both affirming and encouraging certain convictions of superiority and supremacy we either hold dear, feel we should hold dear, or are afraid not to hold dear.

But, again, what are the roots of this canard that we are the "greatest"?

I think it fair and accurate to say that the roots of our Great Self-Image are a curious blend of religious and political elements. Yet, while we obviously have to share these elements in order for Gephardt's superlative falsity to resonate within us, naming those elements is not all that easy or simple. We can sometimes see rather deeply into the water—if we are calm enough, if the water is clear enough—but it's awfully hard to breathe down there. These mythic elements lurk largely below the threshold of consciousness, are made up of things we mostly take for granted, and are lubricated and greased by media oozing "news" and entertainment.[II]

Steven Newcomb, a Visiting Professor in Legal Studies at the University of Massachusetts, Amherst, believes he has located where some of these mythic emblems and submerged dynamics are embedded in an 1823 U.S. Supreme Court ruling. That ruling is called *Johnson v. McIntosh* and, according to Newcomb in "The legacy of religious racism in U.S. Indian law," printed in the April 24, 2002, issue of *Indian Country Today*, it is "based on a religiously racist viewpoint that white Christians are superior to heathen Indians."[1]

This "viewpoint," says Newcomb, solidified in the Johnson ruling, "serves as the legal and political framework within which the United States interprets our fundamental right of self-determination as well as all Indian treaties." The Supreme Court "drew on the Right of Christian Discovery found in Vatican papal bulls of the fifteenth century and in English charters, all of which were premised on the view that all non-Christian lands throughout the world were destined to be taken over by Christians." There's more:

> The Right of Christian Discovery comes directly from the Old Testament of the Bible and is based on the story of the "covenant" between the deity of the Old Testament and the so-called "chosen people." The covenant was based on land. The land "promised" to the Hebrews by the Old Testament deity was already inhabited by the Canaanites, whom the Hebrews were commanded to dispossess. Eventually, this biblical story was transferred to the Americas. As the British international law scholar Henry Sumner Maine put the matter: "In North America, where the discoverers or new colonists were chiefly English, the Indians inhabiting that continent were compared almost universally to the Canaanites of the Old Testament."
>
> Eventually, the story of the divine covenant between the Old Testament deity and the "chosen people" became part of the cultural and linguistic fabric of the United States. It is this biblical perspective that the U.S. Supreme Court inserted into the Johnson decision 180 years ago.[2]

II. For a gripping portrayal of how submerged social dynamics influence crucial life choices, see the novel by Vietnam veteran Tim O'Brien, *The Things They Carried*, especially the Rainy River section, pages 46 through 61, where "Tim" agonizes over whether to go to Canada.

The Superlative Proportions of Our Self-Inflation

Can we begin to see here what "mythology" means? Something we believe to be true without quite knowing why we believe it or exactly how it came to be the underlying fabric of our conviction? To say we are "the greatest" certainly correlates with an ancient Hebrew assertion every Christian child learns in Sunday school—namely, the idea of a "chosen people." To say we are the Greatest Nation that's ever existed in history is certainly to assert implicitly if not explicitly that we are a chosen people or a specially favored people, and even that we have a "chosen" system of governance, of economics, of education, of righteous wisdom, even to the point where the land itself is "chosen." In the early nineteenth century, there already was the pervasive idea of a "Manifest Destiny."[III]

To suggest or assert in conventional company that the United States is not now and never has been The Greatest Country In The History Of The World is tantamount to treason. That is, our mythic self-inflation has reached such superlative proportions that a contrary view, perhaps even the expression of modest uncertainty, is perceived as deadly heresy: not only a political crime, but also a religious sacrilege. And that is to say, our understanding of "country" is invested with unmistakable religious significance and mythological overtones. We are not only the Greatest Country we are also the Best Country. We are instruments of God's redemptive plan. That is, we are Good in contrast to the remainder of the world, which exists in various shades of grey-to-black Evil.

The Greatest Country is also the Good Country, even God's Country: the most moral, the most ethical, the most democratic, the most sanitary, the most productive, the most generous, the most religious—the most modest? Woops! Let's not go *too* far here.

To "powerfully challenge and ultimately liberate ourselves from the legacy of religious racism," in the words of Steven Newcomb, involves nothing less than confronting our dependence on untenable historical and unethical biblical mythology, and withdrawing from that dependence: withdrawing by means of confession and repentance, not by means of sneaky evasion. That is to say, we have certainly entered the terrain of mythological crisis. In the struggle between the hubris of "divine covenant" and the humility of the kingdom of God, we have overwhelmingly chosen to go with the triumphant righteous violence of covenant. The mythological cables that bind us to the patterns and precedents of destructive civility

---

III. For an exacting elaboration of Newcomb's assertions regarding "covenant" and "chosen people," see W. Michael Slattery's *Jesus the Warrior?* Slattery's section in Chapter 2, called "The Ban in Hebrew Scriptures," is especially apt.

are still intact, just behind the decorative emblems of our hubris. It's time for these cables to be cut—the umbilicals that bind us to a mythological womb that no longer nourishes but poisons.

Notes

1. Newcomb, "Legacy," A5.
2. Newcomb, "Legacy," A5.

18

Two Losers and an Icon

THE LOSERS WERE PATRICK Buchanan and Ralph Nader. What they lost, of course, was the presidential election in the year 2000. Their books are, respectively, *The Death of the West: How Dying Populations and Immigrant Invasions Imperil Our Country and Civilization* and *Crashing the Party*, titles indicating, again respectively, the backward-looking racist nostalgia of Buchanan and the forward-looking empirical brashness of Nader.

The icon is Arnold J. Toynbee, British historian extraordinary. The book is *Civilization on Trial*, over sixty years old, but better and more relevant in regard to our present condition and future plight than Buchanan's drippy nostalgia or Nader's combative brashness.

First Buchanan. *The Nation* ran a review, by Philip Klinker, in its March 11, 2002, issue. The cover headline read "Buchanan's Aryan Nation" and the head for the review proper was "The Base Camp of Christendom." It was not a friendly review, and the headlines already point toward Buchanan's theme: nonwhite "races" are outbreeding whites and Western Civilization is about to be swamped by darker skin. Socialism, Marxist culture wars, immigration, and feminism (with its infertile or nonchildbearing women) are all implicated in this demographic disaster. This is a *revolution*, according to Buchanan, that has gone about "dethroning our God, vandalizing our temples, altering our beliefs, and capturing the young." Moreover, this revolution has "captured all the nations of the West." And, "Now that all the Western empires are gone, Western man, relieved of his duty to civilize and Christianize mankind, reveling in luxury in our age of self-indulgence, seems to have lost his will to live and reconciled himself to his impending death."[1] That's Buchanan from his Introduction, just for openers. In the final chapter, "A House Divided," the following remark starts a paragraph: "But if Christianity gave birth to the West and undergirds its moral and political order, can the West survive the death of Christianity?"[2]

It hardly needs to be said that last sentence is fraught with all manner of questionable assertions, not the least of which are depictions of Christianity as mother of the West and undergirding the *political* order of the West. Jesus, I think, would be astonished to discover himself the egg of Christian civilization and the beam upholding the political order. Buchanan's religion is a high mass of sacred empire, larded with a pseudo-humility called "the white man's burden"—our "duty to civilize and Christianize mankind." He's an overbearing priest of white male presumption and civilized affectation, the great wounded warrior staggering with cultural PTSD in the debris of our demographic disaster, John Wayne of the country club in a heroic posture of bleeding-heart conservatism.

As a serious contribution to critical thinking, *The Death of the West* is largely sentimental mythology. But as a statement—perhaps even a manifesto—of potential reactionary vindictiveness, it needs to be taken quite seriously. (We should never presume that sentimentality is free of viciousness.) Buchanan's political mythology has in it a religious mythology that surveys the world (especially the world of darker skin) as other, as enemy, as territory to be conquered for God and Country. It is the stuff of which the "clash of civilizations" is made. It is confirmation of Steven Newcomb's delving into the depths of religious myth that's not only sexist but also racist.

If Buchanan is a charter member of the Old Boys Club, waxing nostalgic over failed empire, Nader is the perpetual Boy Scout Policy Wonk, bursting with enthusiasm for change. If Buchanan blows up a giant windbag of mythology and calls it history, Nader's language carries an abstract scientific precision as he gets impassioned about " noncommercial values," "public enlightenment and civic participation," " a working, deliberative democracy," "patterns of injustice," etc.[3] (That's just from the opening page of Nader's Preface.)

I hugely distrust Buchanan but really love Nader. I believe Nader was the *only* "major" candidate in the 2000 presidential race. He had greater knowledge and deeper understanding of how the system works than all the other candidates put together. But he's such a damned Boy Scout. He knows everything about everything, but he needs twenty years of silence in a Buddhist monastery just to get out of his head. I mean, his head is wonderful, simply wonderful; but one senses in him an intellectual disembodiment that makes citizens, voters in the voting booth, just forget him, or at least not take him seriously. George W. Bush's youthful drunkenness made for a popular emotional rapport, even as his actual policies simply hammered the "common" people. Nader is unequivocally *for* the "common" people—with a dedication and an enthusiasm that has virtually no modern parallel—but as a political personality he seems incapable of setting on fire more than a small slice of the electorate: the bright and idealistic students, some union reformers, the professional muckrakers (but only a certain proportion, for lots of these folks held their noses and voted for Gore and then Kerry), a number of activist farmers and farm workers.

The spirituality in Buchanan is lugubrious, sentimental, racist, sexist, and deadly. Nader is politically correct—truly—but he lacks emotional opulence. His spirituality is so relentlessly abstract and intellectual. In a very real way, Nader is a prophet, and it can be said he suffers the fate of all prophets: everybody knows who he is, hardly anybody takes him seriously.

Although Nader can certainly turn a phrase ("Are we a society stuck in traffic?" or "Become good ancestors"), the bulk of his language is, well, bulky. Here are two clusters of sentences from his final chapter, "Looking Ahead":

> The function of democratic politics is to put forth the forms of societal action that arise from a sense of common pursuits concerning matters of common concern. In a phrase, people to people, all of us pulling our oars in the boat of life where we all find ourselves. The task of an authentic democracy is to make sure that these journeys are accomplished with the informed consent of the governed in a vibrant civil society.
>
> . . . It is character and personality that spell the steady sense of commitment, that give recognition to others in similar endeavors, that enable growth and development of civic skills, perspectives, and frames of reference, that provide the necessary pauses for reevaluation, for improving strategies and modes of self-renewal, for keeping alert and alive the public's imagination of life's possibilities for human beings everywhere.[4]

There's nothing whatsoever *wrong* with these sentiments. They're just so painfully abstract and ungrounded. Reading them, you have to fight off sleep.

Let me point to a place where I believe we can see this problem at work. On the opening page of his final chapter, Nader gives us an example

The Kingdom of God Is Green

of how the "personal experiences of tragedy and mistreatment [can] drive our civic self-respect to engage the necessities of corrective action and not stand by idly rationalizing our futility." The illustration derives from Nader's own experience when, "As a university student one summer, I had occasion to see and feel a little of what migrant farm workers endure every day. I never forgot that those who do the backbreaking work to harvest our foods are paid the least, treated the worst, and harmed the most. That motivated me to speak, write, and act on behalf of these laborers."[5]

OK. Let's look at this picture: migrant farm workers who harvest our foods are paid the least, treated the worst, and harmed the most. And Nader is motivated to speak, write, and act on their behalf. The question is not whether Nader is misrepresenting either the nature of migrant farm labor or his efforts in behalf of migrants. He's not misrepresenting either. The question—if we care to ask it—is a hugely historical and cultural one: *Why* do workers who harvest our foods earn the least, get treated the worst, and suffer the most?

Let me relate Nader's grasp of agricultural issues by (so far as I can tell) the most expressive remark in *Crashing the Party*. It comes in the middle of an account of his attending a Farm Aid rally in Bristow, Virginia. Here's the statement: "By breaking the stranglehold that agribusiness has on the small farm economy, we could increase the share of the food dollar received by the farmer and facilitate a degree of price and quality competition for consumers."[6]

On one level, there's absolutely nothing (from my perspective) to dispute in that assertion. It's politically correct and economically true. Politically and economically, it's right on the money; culturally and spiritually, it's a total wash. Should we care for small farmers because they're hammered *businessmen*? Why? The only argument Nader proposes is that democracy of necessity needs an independent citizenry of largely independent means in order to function adequately.

Once again, I agree. But I find Nader's analysis not nearly deep enough, not delving into the historical, cultural, or spiritual underpinnings of how agriculture got brutalized and why it is so systematically undervalued as a way of life. In the end, an independent citizenry is neither achieved nor sustained by economic formulation or "civic participation" abstraction. It is our creaturely and cultural engagement in Creation that is at issue, and the language for this, implicitly and explicitly, is of necessity rooted in the wholeness and amplitude of our spiritual groundedness.[I]

I. I say in the paragraph above that agriculture got brutalized and has suffered

Is Ralph Nader a prophet for *idealism*? I'm sorry, but as true, honest, and honorable as Nader is, as his message and analyses are, an idealism on behalf of an abstraction just does not cut the mustard. It can be (and in Nader's case, usually is) right on every single issue, and still not be capable of seriously challenging the civilized monstrosity that steadily devours the world and corrupts every culture it penetrates. From the Christian perspective, the main inoculation we have against the epidemic of civilized infection is the yeast in the kingdom of God. But this allopathic remedy has been practically neglected and theologically discredited by centuries of formal Christian quackery. If the kingdom of God is too radical for the "radicals," what's left?

It's time for the icon, Mister History himself—Arnold J. Toynbee. The book is *Civilization on Trial*, copyright 1948. We will nestle in with a comforting quotation having to do with the "philosophical contemporaneity of all civilizations"—just a little something by which to get our temporal bearings:

> On the time-scale now unfolded by geology and cosmogony, the five or six thousand years that had elapsed since the first emergence of representatives of the species of human society that we label 'civilizations' were an infinitesimally brief span of time compared to the age up to date, of the human race, of life on this planet, of the planet itself, of our solar system, of the galaxy of which it is one grain of dust, or of the immensely vaster and older sum total of the stellar cosmos. By comparison with these orders of temporal magnitude, civilizations that had emerged in the second millennium B.C. (like the Graeco-Roman), in the fourth millennium B.C. (like the Ancient Egyptian), and in the first millennium of the Christian era (like our own) were one another's contemporaries indeed.
>
> Thus history, in the sense of the histories of the human societies called civilizations, revealed itself as a sheaf of parallel, contemporary, and recent essays in a new enterprise: a score of attempts, up to date, to transcend the level of primitive human life at which man, after having become himself, had apparently

---

undervaluing as a way of life. All true, as far as I can see. But under this layer of recognition there's more to be conversant with or in search of, namely our noncivilized psychodynamic heritage. If what we now call America had been left to a slow (but not necessarily conflict-free)integration of European peasants and native indigenous, the ensuing transformation would have been so unlike our instantaneous globalized electronic utopia that we wouldn't recognize it. Always in a hurry, utopia pulses with aggressive, restless energy and has been driving eutopia into worldwide destitution and woe.

> lain torpid for some hundreds of thousands of years—and was still, in our day, so lying in out-of-the-way places like New Guinea, Tierra del Fuego, and the northeastern extremity of Siberia, where such primitive human communities had not yet been pounced upon and either exterminated or assimilated by the aggressive pioneers of other societies that, unlike these sluggards, had now, though this only recently, got on the move again.[7]

Well, I'm not sure how comforting it is to be told that human life before civilization was "torpid" or that the fate of some "sluggards" was to be pounced upon, assimilated, or—let's be frank here—*exterminated*. But Mr. Toynbee is not a man to wince at disaster. He tells us that:

> A majority of the score of civilizations known to us appears to have broken down already, and a majority of this majority has trodden to the end the downward path that terminates in dissolution.
>
> Our *post mortem* examination of dead civilization does not enable us to cast the horoscope of our own civilization or of any other that is still alive. *Pace* Spengler, there seems to be no reason why a succession of stimulating challenges should not be met by a succession of victorious responses *ad infinitum*. On the other hand, when we make an empirical comparative study of the paths which the dead civilizations have respectively traveled from breakdown to dissolution, we do here seem to find a certain measure of Spenglerian uniformity, and this, after all, is not surprising. Since breakdown means loss of control, this in turn means the lapse of freedom into automatism, and, whereas free acts are infinitely variable and utterly unpredictable, automatic processes are apt to be uniform and regular.
>
> Briefly stated, the regular pattern of social disintegration is a schism of the disintegrating society into a recalcitrant proletariat and a less and less effectively dominant minority. The process of disintegration does not proceed evenly; it jolts along in alternating spasms of rout, rally, and rout. In the last rally but one, the dominant minority succeeds in temporarily arresting the society's lethal self-laceration by imposing on it the peace of a universal state. Within the framework of the dominant minority's universal state the proletariat creates a universal church, and after the next rout, in which the disintegrating civilization finally dissolves, the universal church may live on to become the chrysalis from which a new civilization eventually emerges.[8]

Toynbee's language is both evasive and prejudicial. What he calls a "dominant minority" is invariably an *aristocracy* imposing a regime of economic extraction and political control. When this dominant minority loses control—watch this language—there is a "lapse of freedom into automatism." This implies, of course, that under the rule of the "dominant minority" the "recalcitrant proletariat" was free ("recalcitrant" means stubbornly resisting authority), but, when the civilized system begins to crumble, "automatic processes" are apt to arise. Why? Scrape away the euphemisms and here's what we get: When the system of control and exploitation begins to break down, those who run the system, in order to maintain control, impose harsher controls.

According to Toynbee, it took several centuries for the Roman Empire to really go to pieces; and when it did, the "true beneficiary of the contemporary Roman Peace was the Christian Church":

> The Church seized this opportunity to strike root and spread; it was stimulated by persecution until the Empire, having failed to crush it, decided, instead, to take it into partnership. And, when even this reinforcement failed to save Empire from destruction, the Church took over Empire's heritage.[9]

That there might be a fundamental contradiction and implacable conflict between "Empire's heritage" and the teachings of Jesus goes unremarked by Toynbee. Or are we to presume that the "Christian Church" had already metaphysically evolved beyond the primitive torpor of its first-century peasant origins? Or perhaps it had just been pounced upon?

So, if the "universal church" that arose from the breakdown of ancient Rome was the Roman Catholic Church, who or what are the leading candidates for the "chrysalis" when the American system goes down? Is it militant Islam? Meditative Buddhism? Aspiring feminism? Could it be Joachim's Age of the Holy Spirit manifesting itself as the Age of the Daughter?

On page 22, Toynbee talks about the competition between Communism and Capitalism, describing them as each other's "rival witch-doctor," and then he says this:

> Yet the fact that our adversary threatens us by showing up our defects, rather than by forcibly suppressing our virtues, is proof that the challenge he presents to us comes ultimately not from him, but from ourselves. It comes, in fact, from that recent huge increase in Western man's technological command over non-human nature—which was just what gave our fathers the

confidence to delude themselves into imagining that, for them, history was comfortably over. Through these triumphs of clockwork the Western middle class has produced three undesigned results—unprecedented in history—whose cumulative impetus has set Juggernaut's car rolling again with a vengeance. Our Western 'know-how' has unified the whole world in the literal sense of the whole habitable and traversable surface of the globe; and it has inflamed the institutions of War and Class, which are the two congenital diseases of civilization, into utterly fatal maladies. This trio of unintentional achievements presents us with a challenge that is formidable indeed.

War and Class have been with us ever since the first civilizations emerged above the level of primitive human life some five or six thousand years ago, and they have always been serious complaints. Of the twenty or so civilizations known to modern Western historians, all except our own appear to be dead or moribund, and, when we diagnose each case, *in extremis* or *post mortem*, we invariably find that the cause of death has been either War or Class or some combination of the two. To date, these two plagues have been deadly enough, in partnership, to kill off nineteen out of twenty representatives of this recently evolved species of human society; but, up to now, the deadliness of these scourges has had a saving limit. While they have been able to destroy individual specimens, they have failed to destroy the species itself. Civilizations have come and gone, but Civilization (with a big 'C') has succeeded, each time, in re-incarnating itself in fresh exemplars of the type; for, immense though the social ravages of War and Class have been, they have not ever yet been all-embracing. . . .

Why cannot civilization go on shambling along, from failure to failure, in the painful, degrading, but not utterly suicidal way in which it has kept going for the first few thousand years of its existence? The answer lies in the recent technological inventions of the modern Western middle class. These gadgets for harnessing the physical forces of non-human nature have left human nature unchanged. The institutions of War and Class are social reflexions of the seamy side of human nature—or what the theologians call original sin—in the kind of society that we call civilization. . . . Class has now become capable of irrevocably disintegrating Society, and War of annihilating the entire human race. . . . We are thus confronted with a challenge that our predecessors never had to face: We have to abolish War and Class—and abolish them now—under pain, if we flinch or fail, of seeing them win a victory over man which, this time, would be conclusive and definitive.[10]

In other words, Civilization made universal, by a combination of intensive global conquests and penetrating technological instrumentalities, magnifies Class and War into such total monsters that the result is bound to be "conclusive and definitive"—that is, catastrophic. Toynbee then asks how, exactly, the "evil of class" has been "heightened by technology?" His answer is that an elevated standard of living does not "cure" the demand for social justice; the "unequal distribution of this world's goods between a privileged minority and an underprivileged majority has been transformed from an unavoidable evil into an intolerable injustice by the latest technological inventions of Western man."[11]

We are circling in on a hugely contentious issue in the center of which the kingdom of God lies waiting. Here again is Toynbee:

> When we admire aesthetically the marvelous masonry and architecture of the Great Pyramid or the exquisite furniture and jewelry of Tut-ankh-Amen's tomb, there is a conflict in our hearts between our pride and pleasure in such triumphs of human art and our moral condemnation of the human price at which these triumphs have been bought: the hard labour unjustly imposed on the many to produce the fine flowers of civilization for the exclusive enjoyment of a few who reap where they have not sown. During these last five or six thousand years, the masters of the civilizations have robbed their slaves of their share in the fruits of society's corporate labours as cold-bloodedly as we rob our bees of their honey. The moral ugliness of the unjust act mars the aesthetic beauty of the artistic result; yet, up till now, the few favoured beneficiaries of civilization have had one obvious common-sense plea to put forward in their own defence.
>
> It has been a choice, they have been able to plead, between fruits of civilization for the few and no fruits at all. . . . In indulging myself at your expense, I am in some sense serving as a kind of trustee for all future generations of the whole human race. This plea is a plausible one, even in our technologically go-ahead Western world, down to the eighteenth century inclusive, but our unprecedented technological progress in the last hundred and fifty years has made the same plea invalid to-day. In a society that has discovered the 'know-how' of Amalthea's cornucopia, the always ugly inequality in the distribution of this world's goods, in ceasing to be a practical necessity, has become a moral enormity.[12]

The Kingdom of God Is Green

Rather than permit Toynbee's allusion to "the 'know-how' of Amalthea's cornucopia" to slip away without comment, as a mere literary irrelevancy, let's just note that, in early Greek mythology Amalthea is (perhaps as a goat) the foster mother of Zeus, and her broken horn, filled by Zeus, was (in the words of *Encyclopaedia Britannica*) the "symbol of inexhaustible riches and plenty, and became the attribute of various divinities and of rivers of fertilizers of the land."[13]

In other words, Toynbee, as historian, is alluding, via elusive mythological metaphor and opaque sexual symbolism, to the agricultural origins of civilization and, simultaneously, implying that "our unprecedented technological progress" has put some know-how into that cornucopia. This literary allusion is about as close as Toynbee—the great historian!—gets to acknowledging the agricultural basis or undergirding of civilization. I offer this *fact* as a sample of evidence indicating that the great academics and scholars, Toynbee among them, have an uncontrolled love affair with "civilization." Do I exaggerate? Here is Toynbee from his fourth chapter, "The Graeco-Roman Civilization":

> In human terms, how are we to describe the Greek civilization, or our own Western civilization, or any other of the ten or twenty civilizations which we can count up on our fingers? In human terms, I should say that each of these civilizations is, while in action, a distinctive attempt at a single great common human enterprise, or, when it is seen in retrospect, after the action is over, it is a distinctive instance of a single great common human experience. This enterprise or experience is an effort to perform an act of creation. In each of these civilizations, mankind, I think, is trying to rise above mere humanity—above primitive humanity, that is—toward some higher kind of spiritual life. One cannot depict the goal because it has never been reached—or, rather, I should say that it has never been reached by any human society. It has, perhaps, been reached by individual men and women. At least, I can think of certain saints and sages who seem to me, in their personal lives, to have reached the goal, at least in so far as I myself am able to conceive what the goal may be like. But if there have been a few transfigured men and women, there has never been such a thing as a civilized society. Civilization, as we know it, is a movement and not a condition, a voyage and not a harbour. No known civilization has ever reached the goal of civilization yet. There has never been a communion of saints on earth.[14]

"Love affair" is, perhaps, not the right term. Here and unmistakably, Toynbee expresses the very thing I allege—namely, that civilization asserts a *spiritual* function in the domestication of humanity. Look straight at this language: "act of creation," "trying to rise above mere humanity" (and then, immediately, the qualifier: above *primitive* humanity), "toward some higher kind of spiritual life." Although the goal may have been reached by "a few transfigured men and women," civilization is, really, a voyage toward "a communion of saints on earth." The terms shift back and forth between the secular and the sacred, they overlap and interpenetrate: civilization's real purpose is to lift us up and out of mere (primitive) humanity and into the elusive realm of the spiritual.

Let's skip back a moment to Nader's *Crashing the Party* for a statement (I presume it is true) on the contemporary scope of the "ugly inequality" Toynbee referred to earlier: "When three hundred of the richest people on Earth have wealth equal to the bottom three billion people on Earth, extreme affluence is built on the backs of extreme mass poverty."[15] That's on page 314 for those who want to check it out.

The economic image one derives from this wealth concentration and income disparity is no different whatsoever than Marcus Borg's "old-fashioned oilcans" with their "needlelike spout rising vertically from the base": an image Borg used, as we saw in a previous essay, to describe wealth concentration and income disparity in first-century Palestine!

So I think we have to question whether those who served "as a kind of trustee for all future generations of the whole human race" really have a "plausible plea," and, furthermore and more deeply, whether the preindustrial unequal distribution of goods for the "underprivileged majority" was, as Toynbee so casually alleges, *an unavoidable evil*—fruits for the few or no fruits at all—and, more deeply yet, whether this fruit-for-the-few-or-no-fruit-at-all can or should be *spiritualized* and therefore sanctified as, at some level, the will of God.

We can, perhaps, zero in on the veracity of this plea by asking a related question: whether War and Class are, in Toynbee's terminology, the "two congenital diseases" of civilization. The ambiguity is in the relationship of these "diseases" to civilization: are they extraneous foreign bodies that can be medically attacked and overcome with moral, ethical, or spiritual antibodies? Or are they so inherently part of civilization's very structure and makeup that "curing" the disease will kill the patient?[II]

II. "Congenital," according to *Webster's*, means "existing as such at birth" and "resulting from one's heredity or prenatal environment." But Toynbee also employs the

The Kingdom of God Is Green

In my estimation, the answer to this question of "diseases" lies, in large part, in how we assess the creative potential of the early agrarian village—Amalthea's cornucopia. Was it capable, on the strength of its own cultural creativity, of evolving into "civilization"—that is, the "voyage" into literacy and the arts, technological inventiveness, peaceful geopolitical expansion—but on the basis of stewardship and sharing rather than militarism and slavery? Or did the "torpid" abundance of the early agrarian village "need" the focused and fierce energy of ruthless armed men in order to force the village's expropriated "surplus" into elite civilization: an "unavoidable evil"? Did the sluggards need to be pounced upon? Fruits for the few or no fruits at all?

To some large but immeasurable extent, this is also a question of how we view "human nature." Are we inherently creative? Or are we inherently slothful? Can "we" do it on our own? Or do "we" (the many) need a "them" (a dominant minority) to whip us into creative productivity? (And, if so, are the "them" necessarily strong men with weapons and a willingness to use them ... on *us*?)

And, beyond the "human nature" level, it is a matter of how we perceive Spirit: are we wretched creatures laboring almost hopelessly for salvation under a stern and maybe even ruthlessly divine taskmaster who will select the rare few for heavenly keeping and sweep the remainder of us into the fiery pit of hell? Or are we blessed creatures helped along in every possible way by a Creator Spirit who, inexplicably, loves us, as it were, without end? And if this loving Spirit is the same yesterday, today and tomorrow, then we can't very well say a harsh God permitted, encouraged, or sanctioned a dreadful system of exploitation lasting five or six thousand years, but now "He" has begun to mellow out because of technological know-how.

---

word "unintentional" in the sentence immediately following the one with "congenital" in it. I realize there is ambiguity here: does he mean the "diseases" were unintentional or that their inflamed globalization was unintentional? Or both? I cannot possibly see how he could mean that Class and War were (or are) unintentional. Such an assertion is simply idiotic, for predatory greed obviously has purpose and is self-evidently intentional. But if one grants that Class and War were (and are) intentional, and if the purpose of War is primarily to expand the realm of political influence, and if control of wealth is the desire of Class, then globalization over time was inevitable by virtue of endemic aggression. And, of course, we have arrived at precisely that point of globalization. But to say this trajectory was unintentional is breathtakingly evasive, for such evasion implies little or no culpability. There is only the thinnest of moral membranes between no culpability and what Gil Bailie calls the enchanting story of founding violence. Either we were (and are) oblivious or God (or the Devil) made us do it. This is not so much idiotic as it is morally sick and ethically repulsive.

I think it fair to say that the kingdom of God approach to these questions does not support the Mad God/Armed Men interpretation. And that means Toynbee's rationalizations—that Class and War are civilization's "diseases" rather than its two ironclad legs, that the brutalization of the majority is an "unavoidable evil"—are just that: evasive rationalizations packaged in a "spiritualized" illusion, all designed to preserve civilizational sanctity. These rationalizations help sandbag and protect founding violence from exposure.

But global disaster is very real to Toynbee. He is, after all, a very good historian. The impending disasters can't be blown off or rationalized away. They aren't mythological. There's no euphemism for annihilation or extinction. The fate of all previous civilizations isn't lost on Mr. Toynbee—and he was writing with Auschwitz and Hiroshima at his back. All this leads him, in his final chapter, "The Meaning of History for the Soul," to examine three distinct points of view: a "purely this-worldly view," a "solely other-worldly view," and "the world a province of the Kingdom of God." He rejects the first two as ultimately unsatisfactory, and he burrows into the third:

> Man [he says] has been a dazzling success in the field of intellect and 'know-how' and a dismal failure in the things of the spirit, and it has been the great tragedy of human life on Earth that this sensational inequality of man's respective achievements in the non-human and in the spiritual sphere should, so far at any rate, have been this way round; for the spiritual side of man's life is of vastly greater importance for man's well-being (even for his material well-being, in the last resort) than is his command over non-human nature.[16]

One cannot help but agree with Toynbee. (One wishes he would agree with himself.) The critical question he then asks is this: "Can there be cumulative progress in the improvement of our social heritage in terms of the spiritual life of mankind," a "conceivable kind of progress," a "cumulative increase in the means of Grace at the disposal of each soul in this world"? Here's his answer—not unlike Thomas Kelly's:

> The actual—and momentous—effect of a cumulative increase in the means of Grace at man's disposal in this world would be to make it possible for human souls, while still in this world, to come to know God better and come to love Him more nearly in His own way.

The Kingdom of God Is Green

> On such a view, this world would not be a spiritual exercise ground beyond the pale of the Kingdom of God; it would be a province of the Kingdom....[17]

And so, in 1948, with the horrific disasters and mass murders of World War II at his back, and the very real possibility of cataclysmic atomic warfare ahead, Arnold J. Toynbee, looking forward, concluded that our only hope would be as a "province of the Kingdom of God"—and that, of course, brings us back to the question of what the "Kingdom" is.

If Class and War are something other than the "diseases" of civilization, if violence (War) is the device by which Class maintains (or tries to maintain) its advantages, and if the industrial revolution democratized exploitation, in part by making extractive violence mechanical and thereby enabling aristocratic Class to become "democratic," then how are we to hope that industrialization provides the cure for the "diseases"? There are two things here, both critically important. First, the industrial revolution has, in fact, served to magnify Class and War, and, thus, it has pushed the destructive capacities of War and the economic "externalities" of Class toward planetary ecocide. Second, the raw fact that industrialization has not eliminated or reduced Class and War, but made them more threatening and more deadly, subverts the "disease" diagnosis. That Class and War are bad "diseases" attached to an otherwise healthy body of civilization is simply and overwhelmingly untrue. And this realization, in turn, causes Toynbee's spiritualization of civilization to crumble.

The grave error of analysis in Toynbee lies in his elevation of civilization into a quasi-sacredness: the means by which we were supposedly lifted out of a subhuman condition, our not-quite-human primitivity. Civilization thus becomes a kind of cultural salvation with spiritual content. Civilization humanizes humanity as it reaches toward a communion of saints on Earth.

This self-serving aristocratic conceit simply ignores, and when it does not ignore it demeans, precivilized and noncivilized culture. It ignores and demeans gatherers, horticulture, agriculture, peasants, and the creative cultural potential of the evolving agrarian village—it spits on Amalthea's cornucopia. It postulates that civilization *accelerated* rather than *retarded* cultural growth when the agrarian village was overpowered and its "surplus" expropriated, when its "sluggards" were pounced upon by the "dominant minority." And, just as it ignores and demeans precivilized or noncivilized life, so it also, on the other side of the equation, idealizes the "trustee" rationalization, the "fruits for the few or no fruits at all," by which Class and War have traditionally been justified by means of a paid priestly and intellectual class.

Yet the best of the intellectual class (Toynbee is an excellent example) sees what Class and War *will do* if they are not abolished. And he of course recognizes, in his final chapter, that the "spiritual side of man's life is of vastly greater importance for man's well-being (even for his material well-being, in the last resort) than is his command over non-human nature." So the great conundrum for the intellectual class becomes how to abolish Class and War while saving Civilization. If Class and War are only "diseases," then it may be possible to find a "cure." I think the Gospel story most apt here is The Rich Young Man (Matthew 19:16–22) who has done, as it were, all the right things—but who is unwilling to let go of his wealth in order to enter the kingdom of God. Toynbee, despite what he says about the importance of "the spiritual side of man's life," wants to believe that Civilization and the kingdom of God are compatible, even that Civilization in its fullest expression is the communion of saints on Earth. The historic truth is that Civilization killed Jesus and will crush to the fullest possible extent any sustained manifestation of the kingdom of God—unless Spirit's fungal yeast is in process of transforming human consciousness. *To abolish Class and War is to abandon Civilization* and to trust (there may be no better word) in the "cumulative increase in the means of Grace"—that is, to dare to enter the kingdom of God devoid of our civilized security blankets. To "repent" means to let go of Class and War. Whether civilization does or does not survive such repentance is interesting but not crucial. Our task, with servanthood and stewardship, is to trust the creative energy of Spirit.[III]

I have attempted here to offer a counter and contrary view to the prevailing conventional understanding of civilization. I propose that civilization is a crime against humanity, a crime against cultural evolution, a

---

III. In the following paragraph, I say civilization is a crime; civilization is a criminal. I believe this is true (though no doubt incredibly complex)—yet true. However, I have a tendency to get a little Manichean—not so much in regard to the folk culture/civilization dynamic that's so obviously true historically, as in regard to the irredeemability of anything civilized. Things do not look good in the world. Disasters are brewing, and that brewing is getting more rigid and more volatile simultaneously: the stuff of explosions. Here's where I say only Spirit can save us, for this arriving Age of the Daughter is the portal through which Spirit is willing to tug us down to Earth: only we have to want it, and we have to want it enough to live it, and the revolution starts when you and I start it. But this also means there are lots of things in civilization that will become eutopianized as folk culture resumes its evolution, this time in a fresh spiritual direction.

crime against agriculture, a crime against Creation, and a crime against the Creator Spirit. I offer for this contrary view four pieces of evidence: the Class and War history of civilization, the present cataclysmic condition of world politics and global ecology as a consequence of civilized globalization, the rise of a protesting, international women's movement as this crisis approaches apocalyptic dimensions, and the vital core of Christian ethics—Jesus' teachings of radical servanthood and radical stewardship in regard to the unfolding of the kingdom of God.

If any of this evidence is deemed inadmissible, superficial, irrelevant, or inadequate, I can only conclude that I have been studying the wrong world or living on the wrong planet. And I beg pardon of those I have detained from football or shopping.

Notes

1. Buchanan, *Death*, 8–9, 10.
2. Buchanan, *Death*, 265.
3. Nader, *Crashing*, xi.
4. Nader, *Crashing*, 313–14, 317–18.
5. Nader, *Crashing*, 311–12.
6. Nader, *Crashing*, 169.
7. Toynbee, *Civilization*, 8–9.
8. Toynbee, *Civilization*, 12–13.
9. Toynbee, *Civilization*, 13–14.
10. Toynbee, *Civilization*, 23–25.
11. Toynbee, *Civilization*, 25–26.
12. Toynbee, *Civilization*, 26–27.
13. *Encyclopaedia Britannica*, Vol. 1, 727.
14. Toynbee, *Civilization*, 55.
15. Nader, *Crashing*, 314.
16. Toynbee, *Civilization*, 262.
17. Toynbee, *Civilization*, 262–63.

19

Ending the Bogeyman Cycle

I AM, NEEDLESS TO say, a person of the Left. The Right, in its best and most ethically coherent sense, may stand for individual liberty—although it's also true that the American Civil Liberties Union (ACLU) is more often aligned with the libertarian Left than with the libertarian Right—while the Left projects a far more inclusive humanitarianism and, in general, a deeper ecological sensitivity. The Right, in recent decades, has been largely swallowed by corporate money dressed in the shabby robes of a worn-out monarchical God. The Right, in other words, is poised for implosion, explosion, or an aristocratic coup.

But I am not glued to a leftist chair, if for no other reason than that the Left has had its full complement of repugnant characters and repulsive doctrines. Stalin may be the biggest icon of Left repugnance. At least he is a prime candidate for such a designation. This fact—the disaster of Soviet communism—disturbed the Western Left to such an extent that its demoralization and contrition have resulted in decades of bewildered uncertainty. There were and are strands of the Left that didn't get swallowed by utopian civility. But we can as a broad generality say that a great deal of the Left's demoralization is due to having had the utopian engineer repeatedly pounded on by both history and nature. Alive in a time of culminating imperial hubris and ecological mayhem, the Left has to rediscover the spiritual depth and earthy power of eutopia.

The Right, however, has done an amazingly comprehensive job of explaining away Hitler and Nazism: they were "pagan" and therefore Devil's spawn. The Left never tried to explain Stalin or Soviet communism as "pagan." The Left had to face the utopian truth. The Right has so far evading facing the truth of its historical corruption. In this sense, the contemporary Left is simply more mature, intellectually and spiritually, than the Right. The Left's demoralized bewilderment is, perhaps paradoxically, a true sign of its relative maturity, while the Right's fierce moral certainty is a smoke screen for its existential panic.

Although the Left's impulse toward sharing cannot and should not be denied, its most comprehensive ideology (if we can call it that) has also taken the civilized inheritance as a given. The Left, generally speaking, has stood for a far more equitable sharing of economic abundance, but it has not been particularly profound in its understanding of civilized dynamics. It has in the main been a big enthusiast for industrialization in virtually all its manifestations and applications. While this enthusiasm has been melting down in the face of environmental limitation and ecological crises—the Left in general is far more aware of ecological issues than is the Right—this melting has yet to really face the raw fact that rampant industrialization is an obvious outgrowth of an underlying civilizational hubris.

The Left (again as a generality) is much more ethically grounded than the Right. (The Right hides its ethical vacuity behind a nearly hysterical assertion of pompous and presumptuous moralism.) But this ethical groundedness of the Left is paradoxically devoid of an explicit spiritual base, whereas the Right vehemently advertises its morality as sitting on—or bolted to—a foundation of religious certainty. That this is so—as regards both Left and Right—is no great mystery, for the Right is desperately in bondage to (in Altizer's words) a "primordial epiphany of God," while the Left has not yet grasped intellectually the "new and absolute immanence" of the new age we're entering. This failure (so far) to understand the breath and depth of transformational immanence has kept the Left, for want of a more profound spirituality, hopelessly attached to utopian materialism—that is, to civilized mythology.

But the Left's utopian materialism is its self-imposed millstone, the cement life jacket it slips into for a little dip in the turbulent waters of history. In short, the Left has recognized and struggled against the Class and War organization of aristocratic civilization, but it has largely failed to recognize (like Arnold Toynbee) that Class and War are not civilization's diseases but the predatory infrastructure components of its very existence. (To say, as Toynbee does, that War and Class are "social reflexions of the seamy side of human nature—or what theologians call original sin" is to evade repentance by saying "I can't help it. It's in my nature. The Devil makes me do it.")

Ending the Bogeyman Cycle

It may never be fully possible for any of us to know completely how we came to be who we are or why we believe what we believe. Our very existence as self-conscious beings is totally inexplicable. Each of us simply is. But insofar as I can discern my leftist roots and inclinations, they are fed by my farm-boy background and by my Christian spirituality. I have very much come to see how agriculture has been under the predatory control of civilization for thousands of years; and I find meaning, not only for my life but for the unfolding of history, in the person of Jesus and in his invitation to the communal banquet in the kingdom of God tent, a banquet at which peasants will have a huge representation.

If Jesus taught simplicity, servanthood, stewardship, and sharing (it's obvious that he did), then these are the values and behaviors that Christians must practice if there's to be any semblance of positive correlation between practice and belief. And insofar as the teachings of Jesus, in broad ethical outline, are like the basic teachings of the world's other great religions and those of other great spiritual teachers, the whole body of such teaching has global reach and universal magnitude. It's love, compassion, forgiveness, honesty, truth-telling, and sharing. Nobody whose ethics derives from one of the world's great religions or great spiritual teachers can ever say that simplicity, servanthood, sharing, and stewardship are not required values and obligatory behaviors. Cook the spiritual books in any religious kitchen in the world, but you've still got these four basic food groups. The Left is the (largely) nonreligious inheritor of these recipes. (When I say "required values" and "obligatory behaviors," I mean this far less in a moralistic sense, as if we were to simply conform to some pietistic code, than as values and behaviors that, by consistent practice, open to us a far fuller dimension of life that can only be accessed via humility, vulnerability, and reverence.)

As I grow older, the central importance of these ethical assertions strikes me increasingly as a no-brainer. At some easily recognized level, we *know* they're true. We know that compassion trumps spite or indifference. And that of course leads us to a question that should only be amusingly theoretical but is, unfortunately, grimly critical: if simplicity, servanthood, stewardship, and sharing are so basic, central, and fundamental to all the world's great religions and great teachers, why do theologians and so-called religious persons condone inordinate wealth accumulation, economic greed, systemic selfishness, national hoarding, and state violence? Why, after thousands of years, are we even having this discussion? Why do we continue not only to have war but war, war, and more war?

The Kingdom of God Is Green

The essential appeal of the Left is its universal humanitarianism. Many on the Left (I'm looking at the September 23, 2002, issue of *The Nation*, with its larger-than-usual bulk devoted to exploring the "reverberations" from September 11, 2001) supported, and still support, the Bush administration's "War on Terror," and even a prospective U.S. invasion of Iraq, because Islamic fundamentalism is equated to fascism, and fascism is to be stamped out, smashed flat, and if possible destroyed without hesitation or delay. A lot of these leftists apparently are atheists, some militantly so. What this seems to mean is that, in their view, there is no God, no Creator Spirit, working historically or biologically in behalf of ethically enlightened human behavior. Enlightened human values, say these militant atheists, are arrived at through civilized refinement, through hard science and critical thinking, and by the shedding of all atavisms associated with conventional religious superstitions, chief of which is God the Bogeyman.

Here we come to a place of rather extreme delicacy. And, like most places of ecological fragility, it is easily trampled into tourist trash by sluggardly bovine traffic, perhaps including the stomping around of my own cloven hoofs.

It's not hard to agree with militant atheists on the Left that God the Bogeyman is the most wicked, or one of the most wicked, of human superstitions. (We might even say—making a very rough pass at being theological here—that such superstition confuses God for the Devil. Or is it the other way around?) God the Mad Dad, God the Irish Old Man with an extremely short drunken fuse, God the Type A personality about to pop a cork, God the Divine Bounty Hunter looking for fresh pelts to throw into Hell, are certainly caricatures; but they are images—packaged a little colorfully here—of a wrathful Father we've been taught to fear with—perhaps for some of us—a few vivid object lessons tossed in for free.

I know one can find this obsessively jealous and moralistic God, find him invoked, in that part of the Bible Christians call the Old Testament. (Recall Steven Newcomb's remarks on "chosen people" as found so brutally frank in Joshua 6 and 7 and analyzed so starkly by Michael Slattery.) There are those who claim such a violent-tempered God is to be found in the Koran. Certainly fundamentalist Christianity is loaded with images of and references to a God of fierce and unremitting judgment: Only a relative handful of people will be saved, the vast remainder will be condemned to eternal torment, and the End of the World is not very far away. Too bad for you. And, no doubt, too bad for me.

Ending the Bogeyman Cycle

The problem, of course, with the terra firma of militant atheistic civility is that if one takes a firm ethical foothold on the evolution of "progressive" civilization, one walks in a pit of very sharp knives, swords, and spears, fully unsheathed. Some atheists will even delight in using the horrific, unsheathed violence of the state to crush such superstitious backwardness as represented by the Taliban or Al Qaeda. Our God will beat the crap out of their God or god.

And so the God who, at one level, is mocked, derided, ridiculed, and scorned—God the Bogeyman—is simultaneously projected as the backwardness to be crushed and invoked as the civilized brute force acting as crusher. We may say we "disbelieve" in this God intellectually, that "He" doesn't exist, but we continue to act in full accord with the psychodynamics of its Augustinian *Realpolitik*. Is there no way out of this Bogeyman cycle?

As is usual with *The Nation*, most of its writers are male. But tucked just past the middle of the current issue is "Whose Security?" by Charlotte Bunch, who points out that "feminists in the United States do not have much impact on US foreign policy, which is military- and corporate-driven,"[1] and that the term "national security" desperately needs to be replaced with "human security":

> But efforts to promote the concept of human security—which emerged out of discussions in which women are active, from the peace movement and the debate over development—were set back by 9/11, with the subsequent resurgence of the masculine warrior discourse. The media have been dominated by male "authority" figures, providing a rude reminder that when it comes to issues of terrorism, war, defense and national security, women, and especially feminists, are still not on the map.
>
> Yet it is women who have been the major target of fundamentalist terrorism, from Algeria to the United States, over the past several decades. And it is mostly feminists who have led the critique of this growing global problem—focusing attention not only on Islamic fundamentalism but on Protestant fundamentalism in the United States, Catholic secret societies like Opus Dei in Latin America, Hindu right-wing fundamentalists in India, and so on.
>
> The events of 9/11 should have generated attempts to address the very real threats to women's human rights posed by fundamentalism, terrorism and armed conflict in many guises. Instead, the occasion was used to demonize the Islamic Other and to justify further militarization of society and curtailment of civil liberties. Growing militarization, often with US support

and arms, has brought an increase in military spending in many other regions, from India and Pakistan to Israel, Colombia and the Philippines.²

Charlotte Bunch goes on to talk about the "rise in the political use of fundamentalism in every religion and region," and how we must "move toward an affirmative vision of peace with human rights and human security at its core, rather than continue to clean up after the endless succession of male-determined crises and conflicts."³

I would like to suggest the following thoughts for consideration. The seemingly endless succession of male-determined crises and conflicts in human history may well be depicted (to invoke a visual image) as the male perimeter of the nomadic band on the move: males, with their (in some respects) stronger and more heavily muscled bodies, always defending the more vulnerable children and the more valuable females. This pattern, or some variation of it, obviously is our cultural inheritance to this very day; only certain distorted features of this pattern were powerfully institutionalized in the foundations of civilization, and the ensuing male triumphalism bent religious understanding so convulsively (achieving, with civilization's rise and consolidation, an all-male authority never before attained) that Spirit itself became identified as overwhelmingly (and overbearingly) Male: both the dangerous but good and protective God of *our* tribe, our kingdom, our country, our empire, our religion, and the dangerous but wicked and threatening God (or god) of *their* tribe, kingdom, country, empire, and religion. Divinity got pumped up as powerfully SuperMale. Both God and Devil are husky brutes—God and Anti-God.

Yet the existence of the world's great religions and their core spiritualities can in no way be reduced to SuperMale Bogeyman superstitions. I would assert that Jesus, Siddhartha, and Lao Tzu—to name only three critical persons—had extraordinary foundational experiences having to do with the significance and mystery of human life and Spirit, and that the ethics deriving from their respective experiences are very much alike, and are very much about sharing, stewardship, servanthood, and simplicity.

The problem with the "application" of ethics is that we tend to want to apply what we think we know without recognizing that our application is dependent on the depth and complexity of our spiritual immersion. Stalin, son of a peasant shoemaker and once a youthful theology student, obviously came to detest the czarist state, wanting to transform that predatory construct into a sharing workers' community, rid of the hated aristocracy. We know what happened—but do we know why?

Ending the Bogeyman Cycle

I don't believe Stalin ever grasped that a workers' civilization is still civilization, that civilization is inherently based on imposed class chasms and immense wealth cleavages, on institutionalized violence and systemic oppression, on organized theft and plunder, on the expropriation of agriculture and contempt for the "backwardness" of the agrarian village. Stalin proved to be a thoroughly civilized man, perhaps nothing more than a new Russian czar who chose to industrialize in a hurry. He is, in a sense, proof that Toynbee's "disease" analysis is both inadequate and false: keep or expand civilization and you can't help but keep or expand Class and War. (Mikhail Gorbachev was far less Stalin's heir than is Vladimir Putin.)

The core problem, the essential obstacle toward arriving at "human security," an "affirmative vision of peace with human rights and human security at its core," is, first, the false projections of dreadful male divinity and, second, the false spirituality of male civilization. If Islamic, Christian, and Jewish fundamentalisms represent the core of the first problem, or at least a big hunk of that core, all those who are so breathlessly and uncritically for civilization are at the center of the second problem.[1]

So what does it take to bring to a close this "endless succession of male-determined crises and conflicts"? Extermination is one possible closure. Restoration of explicit aristocracy may be another, at least insofar as it would radically shrink "democratic" consumerism. The finest route is immersion in Spirit, fully and truly in over our heads. The women's movement suggests that the Midwife who waits to deliver us will be very glad not only to help us be born again, She also will be enormously relieved at our having had our fill of an endless succession of male-determined crises and conflicts: little feral boys who, having bloodied and battered the

---

1. We might say that patriotism (and there is no patriotism without *pater*) is compulsory male bonding. We might even say that it was male bonding at the level of the state that put Jesus to death. Insofar as this bonding was patriotic (see John 11:47–53), we can also say—at least psychoanalytically—that the Father killed the Son because the Son was consistently refusing to behave within the rigid confines of male-bonding patriotism. This suggests that orthodox Christianity is really about two primary figures: a dead Son hanging on a cross and a brooding Father who can neither forgive the Son for his refusal to be patriotic nor cease feeling perpetually guilty—not only about the murder of His Son, but also about two other related matters: His total suppression of the Mother (a subject not to be opened even slightly), and His very own inability to relinquish His righteous, wrathful superego. This also suggests that the rediscovery of the kingdom of God as the primary proclamation of Jesus—the "program" of his (un)patriotic renunciation—announces the end of a religious stasis. The Son's message is resurrecting in a world where wrathful, righteous superego, both in a religious and civilizational sense, has created a crisis of epic global proportions.

The Kingdom of God Is Green

world beyond all past excesses, are finally ready to grow up—and who are spiritually strong enough to tell Dad to grow up, too.

Notes

1. Bunch, "Whose," 36.
2. Bunch, "Whose," 38.
3. Bunch, "Whose," 40.

20

The End of Something Big

CHRISTIAN FUNDAMENTALISM IS PRONE to right-wing domination for a variety of reasons. First, it is antischolarly and anti-intellectual, having no truck with any analysis that leads one outside of or beyond six-day instantaneous creationism. In that respect, fundamentalism simply shuts down the mind. Second, it depends on a totally conventional theism of an all-powerful Father God who seems to reside somewhere outside the universe, but who will save for eternity a relatively small number of true believers and condemn for eternity a much larger number of unbelievers and sinners. Third, this all-powerful Father God is going to intervene directly and massively in world politics when certain signs are at hand and certain preconditions are fulfilled, just as prophecy foretells. Fundamentalism is, among other things, the literalization of this myth, a myth that has (is it really only incidentally?) an amazingly tight correlation to the unrestrained prerogatives of a king (God), the certainty and confidence of an established aristocracy (the eternally saved), and the perpetual hopelessness of an oppressed peasantry (the eternally damned).

Two articles—one in the June 3, 2002, issue of *The Nation* and the other in the August 2002 *Washington Report on Middle East Affairs*—underscore how pervasively strong, how ruthlessly monomaniacal, these convictions can become when mythology is fiercely promoted as true history and impending eschatology. (Of course what we're seeing, at least in part, is an eruption of the apocalyptic in the aftermath of 9/11.) If fundamentalism is an armored strongbox of mythological conviction, the apocalyptic is a time bomb ticking loudly within. Some of us considered such beliefs and convictions a relatively harmless form of religious aneurysm—good for either a little emotional ecstasy on occasion or some colorful theological apoplexy; but we need to reconsider our happy, smug view. Religious apocalypticism is now diddling in world politics in such a way so as to trigger a cerebral vascular incident on an international scale, a political stroke booby trapped with an unimaginable magnitude of ecocidal weaponry.

Well, here are the articles. From *Washington Report* we have Allan C. Brownfeld's "Strange Bedfellows: The Jewish Establishment and the Christian Right," and from *The Nation* we have Deanne Stillman's "Onward, Christian Soldiers." Make of them what you will, reaching either for soothing indigestion tablets or some reviving digitalis.

The smaller headline above Ms. Stillman's article reads: "In Anticipation of the Second Coming, Evangelicals Leap to Israel's Defense." Both articles—Ms. Stillman's and Mr. Brownfeld's—focus on the linkage between Christian fundamentalism and the state of Israel, how the creation of modern Israel was and is understood as a step toward the fulfillment of biblical prophecy, how the Dome of the Rock—the Muslim mosque supposedly on the site of the Jewish Temple Mount—must be destroyed so that a new temple can be built, how Israel must expel the Palestinians and even claim all land between the Euphrates and the Nile so that Armageddon is hastened and the Second Coming of Christ prepared for.

Will the world be blasted? Of course. But Christ will lift the saved into the clouds. This is the "rapture." Here is Deanne Stillman:

> For these evangelicals—and for messianic Jews as well—the approaching doom is not about Saudi oil, the right of return for displaced Palestinians, the war against terrorism, Bush family interest in the Carlyle Group or Ariel Sharon and Yasir Arafat as warriors who can't disengage. For them, the fight for Israel boils down to a particular shrine in Jerusalem called the Temple Mount, where Christ is supposed to return. As it says in Micah 4:1, "In the last days the mountain of the Lord's temple will be established as the chief among the mountains; it will be raised above the hills, and people will stream to it."
>
> This is the heart of the matter, the holy of holies (yes, there are plenty of other holy of holies, but none as holy as this one, which is why Christians weren't all that upset about what went on at the Church of the Nativity, where the thing that was supposed to happen there—the birth of Christ—already happened.) This is the full-on shmegeggy of religious weirdness, an ancient pile of rocks and amens and allahu akhbars, the place

The End of Something Big

that sparked the intifada of the past twenty months. There's a rock here that God asked Abraham to sacrifice his son on and Abraham said OK and then God said I'm only kidding, I just wanted to make sure you adored me. Then Abraham became the father of the Jews and to honor Abraham, a descendant of his named Solomon built a temple on top of the rock. Solomon's enemies trashed the temple. The Romans rebuilt it and, not knowing Abraham or Solomon, named it after Herod. Later, Mohammed came along and said Abraham was the father of the Muslims. One day, while standing on this same rock, Mohammed ascended to heaven. So the Muslims built the Dome of the Rock on top of what was once not one but two Jewish temples.

This presents a major problem for Christians: How can Jesus make his scheduled landing at the Temple Mount if it's beneath a mosque that happens to be under Muslim control? (Such is the politics of Israel that Jews can control a city while Arabs can control a building in it, although Jews control a wall of said building—which happens to be the one-and-only Western Wall.) Solution: Build a third temple on the mount. After all, the Bible—Old and New Testament—says there's supposed to be a third temple. Which is why Jerusalem cannot be divided in half, as Palestinians have suggested, because the half that they want—the one with the Dome of the Rock—would be part of their side. For fundamentalist Christians (and Jews) it's bad enough that Muslims already control it. When Sharon visited the site there in 2000, the place went off because Palestinians took it as a sign that the Israelis were getting ready to dismantle the mosque and restore the Temple Mount, which Christians saw as proof that Christ was around the corner.[1]

The political problem for any Palestinian leader, says Stillman, is the "unspoken view that Palestinians are blocking a divine plan. And so here Jews find themselves: temporarily in lockstep with the conservative Christian world, the same one that stood by and watched them march to the ovens."[2] So it's not that the "conservative Christian world" loves the Jews; it's that certain events in Jerusalem, working through the "chosen people," reveal God's Plan, a plan that includes eternal salvation for conservative Christians.

After informing us of a variety of recent linkages between Christian and Jewish fundamentalists, Allan Brownfeld takes us back in time:

> The roots of Christian Zionism go back to the Protestant Reformation. Before that time, all Western Christians were Catholic

and generally accepted the view taught by St. Augustine and others—that certain biblical passages should be interpreted allegorically not literally. As an example, Jerusalem and Zion were heavenly, other-worldly—open to all of us, and not actual places on earth to be inhabited exclusively by Jews. By the 16th and 17th centuries, however, Christians for the first time were buying Bibles and interpreting Scripture for themselves. In doing so, they began to elevate the concept of Israel—and the Jews—as the key factors in biblical prophecy. Bible-loving Christians came to regard the Old Testament as the only history that mattered in the Middle East. . . .

It is unlikely that many members of the Jewish organizations now embracing the Christian Right understand the motives and theology of their new allies. Do they understand that Jerry Falwell, Pat Robertson, Tom DeLay and the others support Israel's most extreme policies—even the "transfer" of Palestinians from the West Bank—*not* because they seek Middle East peace, but because they are encouraging conflict which, they believe, will hasten the end of the world and the Second Coming of Christ? And what would become of Jews if this scenario occurred? Those who did not become Christians would be condemned to hell while their "allies" were raptured to heaven.

To the extent that U.S. Middle East policy is influenced by such an apocalyptic vision it becomes an instrument which sows discord and makes genuine peace increasingly unlikely. Jewish groups making a theology of embracing every twist and turn in Israeli policy find themselves in a strange alliance with those whose dream is a violent end of the world. It is this dangerous confusion of religion, politics and foreign policy which leads to such strange bedfellows and their current embrace. Such an embrace is likely to bear very bitter fruit indeed.[3]

That there are huge numbers of people, politically acute and financially capable, deeply committed to "encouraging conflict which, they believe, will hasten the end of the world and the Second Coming of Christ," should scare the crap out of us, leading to the defecation of our mythological meconium. This is deadly serious, even with the semihumorous scatological metaphors fully apt. When a huge grouping of mythologically charged people, powerful in their political connections with the most militarily strong nation in the history of the world (this is *not* hyperbole), are openly meddling in international political events in order to achieve otherworldly and calamitous results, we are indeed faced with the end of something big. At its most unimaginably disastrous, this meddling could unfold as the end of the world for many higher life forms on Earth, most certainly including ourselves—not a pretty prospect.

That this is a military-industrial-political psychosis based on a civilized, utopian, and religious meshing of mythologies is starkly obvious; and it will become, I'm afraid, even more obvious in the years ahead. From within these respective mythological fungi there will be no slowing or stopping of growth that I can foresee, but, rather, a continual merging and metastasis of mythological distortion until the entire systemic mass reaches the end of its growth, sags, and disintegrates.

The ecological and cultural cost of this morbidity is beyond reckoning. But yet one has hope, for hope is embedded in life, embedded in Creation. Even those who are mythologically drunk or drugged on the inverted erotic sadism of Armageddon are creatures of Creation; only the hope of "rapture" is so colossally judgmental, secretly selfish and prideful, that it has become an ecological suicide cult.

I think all this requires a careful reconsideration of Thomas J. J. Altizer's bold assertions that "both Paul and the early Church were unable fully or decisively to negate the religious forms of the old history, or to surmount their bondage to the transcendent and primordial epiphany of God," that the church has become "estranged from its own initial proclamation," that the "original heresy was the identification of the Church as the body of Christ," and that when the church "is further conceived as a distinct and particular institution or organism existing within but nevertheless apart from the world, then the body of Christ must inevitably be distinguished from and even opposed to the body of humanity." Compare and contrast what Deanne Stillman and Allan Brownfeld say with this (previously quoted) remark of Altizer's: "Once the Church had claimed to be the body of Christ, it had already set upon the imperialistic path of conquering the world, of bringing the life and movement of the world into submission to the inhuman authority and power of an infinitely distant Creator and Judge." (This certainly challenges, at least implicitly, Brownfeld's assertion that, until the Reformation, Augustine's emphasis on allegory held sway. Remember Steven Newcomb telling us that the 1823 Supreme Court ruling on Indian treaties had part of its root in fifteenth-century papal bulls and, even more ancient, the Old Testament "covenant" between Yahweh and the "chosen people.")

The culminating movement of that attempt at forced submission has arrived, brimming with apocalyptic anticipations and oozing with toxic disasters. Meanwhile, the "kingdom" of God still is waiting. A rounder, more comprehensive, more fully humane and ecological mythology is possible in the spiritual dawn, a mythology less allergic to sunlight. But first, it seems, the righteous priests of male disaster must dance before their god of war and invoke his blessing of destruction.

Notes

1. Stillman, "Onward," 28.
2. Stillman, "Onward," 28.
3. Brownfeld, "Strange," 72.

21

Divine Terrorists

My wife Susanna does not like the sharp tongue I hone in the preceding essay. "Does not like" is something of an understatement. She is right about lots of things and she may well be right about this. But that doesn't, however, mean I will blunt my tongue either now or retroactively. (A writer's words cannot be considered fully winged until a willing publisher has consented to let them fly. Until then, blunting, clipping, pruning, slashing, and hacking are possible. After that it's only regrets.)

It has dawned on me, after rereading my assertion that a significant slice of the Christian Right has become an ecological suicide cult, that these folks (there are, apparently, a fair number of them in the U.S. Congress) have a crude model from which to work. I speak of the nineteen men who hijacked four planes on September 11 with the conscious intent of achieving a small-scale Armageddon in order to provoke a "holy war" against the West—the United States in particular.

The Christian Right claims to be following a divine script—literally—that not only permits but encourages them to deliberately and intensively heighten a "conflict which, they believe, will hasten the end of the world and the Second Coming of Christ," according to Allan Brownfeld. A "holy war" is just a smaller and more localized form of the divine terrorism the Christian Right is seeking openly and officially to provoke, *in order to bring an end to human life on Earth*. Literally. This is, we might say, the end game of apocalyptic globalization.

Are we still too timid to ask who the really big terrorists are? They are the ideological offspring of a mythology that paints the Creator as Head Cosmic Terrorist. The little guys are only following orders. The big guys claim to be hearing voices.

22

The Gendered Feminine and Twenty Centuries of Papa

PATRIOTISM, THE PAPACY, AND the pope are all, etymologically, about papa, about papa's jurisdiction, about what was established and ordained by the forefathers. They're all about allegiance to a mythology, a system, and a theology that is, needless to say, explicitly and exclusively male.

So, do we really want genderless patriotism, genderless war, or a woman pope? Is this really what democracy, women's liberation, and the kingdom of God are about? Equal opportunity to operate traumatic institutions or run the civilized system of War and Class, all of it protected by God the Father?

Neil Douglas-Klotz, an "independent scholar of religious studies, spirituality, and psychology," was interviewed in the Summer 2000 issue of *Sounds True*. Proficient, apparently, in both Aramaic and Hebrew (scholars say Jesus spoke Aramaic), Douglas-Klotz says that "The Aramaic word for 'kingdom,' *malkuta*, is gendered feminine, so it is better translated as 'queendom.'"[1]

If this is true, it's no wonder the kingdom of God has been neglected, avoided, and evaded by twenty centuries of patriotic papas. Inside this "kingdom" is the gendered feminine of God. But, if we say "queendom," shall we also say—instead of the "kingdom of God"—the queendom of the Goddess?

Notes

1. Douglas-Klotz, "Original Prayer," 18.

BOOK TWO

Woman's demand for equality has the power potential of a movement that can change the course of history by adding a new dimension—the feminine dimension—to the conduct of human affairs after thousands of years of male domination. The number of men who recognize that the world is moving towards almost certain disaster is swelling day by day, all over the globe. They concede that humanity's scale of values must be changed at the point where power is generated and policy imposed, and not merely within governments. An unexpected and compelling influence must be brought into the management of human affairs, capable of carrying weight where power takes shape and where decisions are made as to how and where it shall be used. Such an influence can only be the influence of woman because, in a bisexual world, she is man's only peer, even if millennia of history have seen her treated as his inferior. She must, however, become man's equal partner, and not merely his equal in the sense that he accepts her on his own terms in a society tailored to his traditional supremacy. She must become his equal in a partnership of equals. If woman is to play her full role she has no alternative to making this her goal.

—LISA SERGIO,
Jesus and Woman, pages 1–2

Recovering the feminine principle as respect for life in nature and society appears to be the only way forward, for men as well as women, in the North as well as the South. The metaphors and concepts of minds deprived of the feminine principle have been based on seeing nature and women as worthless and passive, and finally as dispensable. These ethnocentric categorisations have been universalized, and with their universalisation has been associated the destruction of nature and the subjugation of women. But this dominant mode of organising the world is today being challenged by the very voices it had silenced. These voices, muted through subjugation, are now quietly but firmly suggesting that the western male has produced only one culture, and that there are other ways of structuring the world.

—VANDANA SHIVA,
Staying Alive: Women, Ecology and Development, page 223

For wherever there is culture and civilization there can also be counterculture and anticivilization.

—JOHN DOMINIC CROSSAN,
Jesus: A Revolutionary Biography, page 117

Asked when the kingdom of God was to come, he gave them this answer, "The coming of the kingdom of God does not admit of observation, and there will be no one to say, 'Look here! Look there!' For, you must know, the kingdom of God is among you."

—LUKE 17:20–21

23

Age of the Daughter

WE LIVE IN THE closing phase of a time in which a lot of people have wished to be rid of superstition and myth. As a society, we have generally been glad to "demythologize" something or other, to step into the bright light of empirical truth and out of the shadows of superstition and myth. But the deeper truth is that we live in a time in which the consequences of accrued empirical truth, largely owned and operated by the "traumatic institutions" of civilization, and therefore also reflective of a "diseased" consciousness and aspiration, have brought us to an apocalyptic bouquet at the swiftly approaching edge of an abyss. Traditional myths are no longer adequate or able to speak prophetically to our circumstance—except by an apocalyptic script that mirrors and reinforces what empirical truth has accomplished or is accomplishing in its control of civilization. Meanwhile, no new myth has emerged or congealed.

We have largely outgrown traditional myth, although fundamentalism clings mightily to its literalistic obsessions, providing a huge anchorage for masculine prerogative in all three Abrahamic religions; and (in a psychologically complex way) myth serves to justify or excuse the aggressiveness of civilized economics. "We had to destroy the village in order to save it."

Because the political space around fundamentalist conviction is so dangerously charged with fierce religious assertion simultaneously with the aggressive hegemony of civilized economics, we are allowed no clear exploration of anything resembling a steady-state future. Such debate is

The Kingdom of God Is Green

not encouraged.[I] We have no mythic Gestalt for that which lies ahead, even though we may be able to point with urgency to this or that critically needed change. Hence there is, amid great global troubles and growing social uneasiness, a general inability—even a reluctance or refusal—to formulate a comprehensive cultural and political alternative. We are, in fact, living in the midst of a politics of denial, and this denial is based on an accrued refusal to look honestly and deeply into inherited myth.

The "War against Terrorism," overtly so in the aftermath of September 11, 2001, has begun to spin the ancient hostilities between and within the Abrahamic religions into epic global proportions. We also have a globalized system of civilized infrastructure that is bringing, as Paul Hawken says (in the epigraph that opens Book One), "every living system" into decline; and into this process of accelerated decline, the three Abrahamic religions are poised for bloody warfare, each of which (in wildly disproportionate numbers) contains nuclear weapons in its arsenal. (Those who claim to stand with the Prince of Peace have far and away the most, the biggest, and the best of the weapons.)

Since human beings act to a large extent according to the channeling of myth, we may safely say that the global crisis we are in, and into which, it seems, we will more deeply plunge, is the crisis of myths we are trapped inside of but that are no longer functional or adequate. They are, in fact, the vehicles *causing* our crises to reach apocalyptic dimension.[II] We are in the painful, bloody, and destructive process of discovering the depth and magnitude of this "present disorder," and that means we are also being forced to outgrow our inherited myths, of being mythically born again. But, perhaps like all mammalian births, this one too will be involuntary, messy, and painful.

What do we mean by "myth"? As a mild generality we may say that myth is an overarching depiction or conception of reality that gives

---

I. Consider how "single-payer" or "Medicare-for-all" was disallowed any consideration during the *Democratic*-controlled process of healthcare reform legislation in the first two years of the Obama administration. Or, as the warnings of climate change grow more alarming, how there is virtually no federal effort to reduce fossil fuel consumption, reverse agribusiness consolidation, encourage clean alternative energy, or build an adequate mass transit system.

II. Even such a conflicted radical/conservative Christian as Reinhold Niebuhr, writing in 1948, recognized that "Western civilization has been the center and source of the world's disorders," a remark made in the opening paragraph (page 103) of an essay entitled "God's Design and the Present Disorder of Civilization" in *Faith and Politics*.

concerted meaning to human purpose and that provides an ethical framework by which to understand history and coordinate human effort in the natural world.[III] Myth is or expresses the numen or presiding spirit around and through culture. Myth tells us what is desirable and undesirable, humane and inhumane, and it does this by means of stories and the inflation of stories. To live without myth is to sink, or to risk sinking, into existential anomie. Without myth, without a rounded, publicly accessible conception of the possible, of what it means to be human, life can easily and quickly become listless and meaningless. And meaninglessness raises both apathy and self-indulgence to the level of social norm or even political principle. Stories tell us who we are and where we've been.

Sensory deprivation of a kind is possible for the mythic needs of the soul. To attempt to live without myth—without stories—is an austere stoicism of rational will. To live by total empiricism is to enter the realm of the robotic. An owner's manual is not a story. It is here where meaninglessness truly threatens. Myths, therefore, even if denied as myths, assert themselves out of the bleak mundane: the "myth of the machine" or the "myth of progress." Indeed, insofar as we yearn for a future of personal fulfillment and ecological wholesomeness, and insofar as we fill that yearning with images of desire, we are delving into the territory of myth: projections of hope or expectation cast upon the future that provide direction and channeling for our energy. When all-embracing cultural myth is shed—usually by being broken into by developments the old myth cannot explain or accommodate—more personal images are formed in lieu of collective myth. Myth is replaced by private fantasy—though in our culture, thanks to capitalist commodification, even "private" fantasy is for sale via electronic communications and video games.

In the West, the Judeo-Christian myth—its primary conception of extraterritorial divinity, its account of creation, its metaphysics of

---

III. In a footnote on the opening page of Chapter 16, "The Superlative Proportions of Our Self-Inflation," I briefly quote from Gil Bailie's *Violence Unveiled*. Bailie says—and I agree—that "myth remembers discretely and selectively," often to provide "an enchanting story of . . . founding violence." That said (and its lesson always needs to be kept in mind), we are creatures who have history, who tell stories. But not all stories are veils for violence. We are, in a sense, no more able to live without myth than we can function without the unconsciousness. And, if we care to utilize Freud's trio of id, ego, and superego, we might say that myth (in its best sense) arises from the id but that the superego in its obsessive lust for moral self-justification is the force that bends myth toward the rationalizations of "founding violence." "God made me do it." Violence at that level is so inescapably wicked that only the blessing of the divine can make it endurable.

salvation, and its otherworldly eschatology—no longer holds credibility for a great many people, although both the cultural ethos and the social institutions of the West were largely (if also ambiguously) shaped in a Judeo-Christian crucible. Our *civilization* derives from Rome, Greece, and the earlier civilizations of the ancient "fertile crescent," and it is the consolidation, the globalization in capitalist form, of a predatory impulse crystallized thousands of years ago when male bandits expropriated the abundance of the agrarian village.

Civilization, precisely because it has never shed its predatory origins (its institutions are "traumatic" because they were forged out of trauma), is the core dynamic in our enlarging crisis. That is, the global spread of the Euro-American capitalist system, following on the heels of a more fragmented European colonialism first disrupted and then broken by two world wars, has ruptured local and regional cultures around the world and turned large numbers of people, in various degrees of antipathy, against the West. The effort on the part of "terrorists" to activate frustrated and angry Moslems in a "holy war" against U.S. economic and military hegemony is a bubbling of this vast antipathy, with the West's post-World War II shoehorning of Israel into Palestine a major thorn in the lion's paw. Religion (or religious identity) thus becomes a mobilizing vehicle for the expression of grievances that could be articulated in other ways.[IV]

But just as civilization, with its congealed male warrior energy, is predicated on violence and armed force, so the three Abrahamic religions have traditionally postulated a mighty male Father as Creator of the universe, a jealous God requiring singular devotion, a God who may inexplicably love His "chosen people" beyond all others but who also hates and is wrathful towards those not chosen, even promising them an eternity of torment in hell. In other words, this is a conflict of rampant maleness, of male *systems*, built on a base of warrior aggression and bandit cover-ups; and from this conflict there appears to be no exit except through the exhaustion or dismantling of male hegemony—and that includes the implosion of Abrahamic myth.

IV. But "other ways"—i.e., more openly democratic ways—have often been blocked or thwarted, as when British and American intervention in Iran in the early 1950s undid Iranian democracy and installed the Shah, who in turn suppressed democratic aspirations and thereby channeled opposition into the last remaining avenue—the mosque.

II

The leading religious and political institution within European civilization, following the decline of Rome, was of course the Catholic Church. The religious hegemony of the Roman church was subsequently broken by the Lutheran, Anabaptist, and Protestant reformations in the early modern period. Within a few centuries the splintered religious ethos was, if not supplanted, then compelled to provide more and more social room for a growing secularization. This secularization correlates closely not merely with the powerful discoveries of science, but also with the rise, spread, and hegemony of capitalist economics on a global scale.

Church and state, religious myth and secular ideology, became self-conscious rivals for the affection and allegiance of the people. (They are also, on occasion, as in much Republican political posturing, each other's defender and apologist.) Political ideology became, in a sense, a narrow form of rationalized myth, especially as "progress" imitated heaven (that is, it "democratized" aristocratic opulence in the form of industrial commercialism) and sapped its otherworldly energy. The religious goal of life in heaven was turned, slowly, steadily, and not without internal contradiction, toward a secular goal of paradise-through-progress. This secular hope fastened itself to the expectation of continual material and technical progress, to economic growth, and to convictions regarding mortal victory over nature and the postponement of death. If science could not defeat death, it could at least frustrate its early arrival and postpone decomposition by means of formaldehyde.

The enlightened secular mind began to see the universe as a huge machine and human industry as the tireless engine of Progress. Progress became our most important "product." Humanity, to some extent, broke free of heaven only to be captured by civilized utopian ideology offering limitless consumerism and ersatz immortality via the mortuary. If heaven represented otherworldly escape from death and evil, progress promised earthly liberation from backwardness and disease, although neither civilized religion nor civilized ideology was particularly interested in a deepened cultural integration with ecological community, and certainly not interested in any sort of unity, for example, with Native American cultures. Both indigenous cultures and the peasantry were in the realms of evil (at worst) or backwardness (at best), and they have paid a very high price for the civilized indulgence in utopian supremacy, with its largely compliant religious little brother who, more often than not, willingly preached the evils of "paganism."

The Kingdom of God Is Green

Religious humanists, partly to break free from the unscientific literalism of established religious creeds and partly to bridge (or try to bridge) the growing chasm between the religious and the secular, "demythologized" a great many religious conceptions and rendered religion an intellectualized assemblage of well-intentioned ethical outlines, moral prescriptions, and literary analysis. The effect was to essentially secularize, although primarily in abstract utopian form, the mythic turf of eschatology, the biggest vision of transformation or of ultimate concern. That is, religious humanists helped take the oxygen out of heaven and transfer it to progress. Religious conservatives, on the other hand, largely in reaction against the liberalizing efforts of humanists, literalized their theological conceptions and fortified religious myth into a take-it-or-leave-it grocery list of precisely worded beliefs. This is the origin and meaning of "fundamentalism."[V]

Liberals deliberately diffused religious myth in the direction of civilized progress; conservatives just as deliberately literalized myth as a kind of desperate fetish. Both endeavors were symptomatic of a stressful time in which religious myth was under severe pressure from an increasingly powerful economic ideology, a constantly modifying technology, a growing scientific skepticism, and a social order radically disrupted by all this "progress." The traditional Judeo-Christian myth was forced either to dissolve or suffer a hardening of the intellectual arteries. Dissolving engendered emotional diffusion and identity loss; hardening brought on intellectual rigidity and an increasingly chiseled self-concept dependent on strict adherence to precisely articulated beliefs. Liberals mostly became apologists for progress while conservatives largely became fiercely judgmental about all the unrighteous sinners (pagans, heathens, infidels, unbelievers, and other backward peoples) whose false convictions, laziness, cultural limitations, and skin color were earning them the eternal wrath of God. Increasingly, God's wrath was also directed toward adherents of Left ideology, toward feminists and homosexuals, and toward anyone who believed evolution plausible or thought abortion a reasonable (even if difficult) choice in some circumstances. In short, God was turning His angry face not only toward conventional heathens and traditional pagans, but now also toward cultural and political liberals.

Science, meanwhile, was propounding a radically new myth (i.e., a cosmic picture) of evolution and natural selection. Although it offered a

---

V. The five "fundamentals" of Christian fundamentalism are: "the inerrancy of Scripture, the divinity of Jesus, the Virgin birth, Jesus' death on the cross as a substitute for our sins, and his physical resurrection and impending return." So says Justo Gonzalez on page 257 in *The Story of Christianity: The Reformation to the Present Day*.

grand vista of the past, complete with monstrous dinosaurs and frigid ice ages, science was limited in what it could offer for the future. Looking to the past, science could become eloquent on the complexities of astronomy, geology, the evolved natural world, and the rich diversity of life; looking forward, it could only postulate a thin probability of progress, with "man" at its apex. Yet this thin probability was tremendously exciting to a great many people. And, since science claimed to be both neutral and objective, no impediments were foreseen in the growth of secular freedom, material abundance, and the conquering of nature. To be civilized was a grand thing, destined to be even grander. Science saw itself as the new and far more competent driver of the vehicle called civilization. Few realized that science, derived heavily from the excessively abstract and sexually typecast conceptions of ancient Greek rationality and from the otherworldly psychology of Christian conviction, was seriously burdened by an implicit reductionism made even more reductionist by its financial captivity and research channeling by state priorities (militarism) and capitalist economics (commodification). Science, especially in its subsidized forms, became—perhaps ironically—far less civilization's competent new driver than its intellectual eunuch soaking in the hot tubs of pampered "think tanks," guided by commercial aggressiveness and a domination complex. This reductionism, priding itself on a new liberation from ignorance and superstition, quickly became positivist and empiric and bred a new intolerance for those who questioned either the means of science or its ends: doubters merely reflected a kind of dismissible timidity derived from anxieties inherent in traditional superstition—reflexively opposed to progress like any unenlightened savage.

Yet myth, to be socially creative, culturally dynamic and ecological sensible, needs to provide a supple, living metaphor by which priorities of value can be broadly ascertained; and it needs to do so in ways that do not require a mind-wrenching belief conversion. Science, for all its promises of technological progress, has not, for the most part, deeply touched the inward yearning for personal transformation or ecological community. Its energy has been captured and channeled by utopian civilization; it has only in a very limited sense been encouraged in a eutopian direction—household solar panels, perhaps. Backyard gardens. But even these things are often understood to be mere lifestyle enhancements for an affluent class playing at living "green." The main thrust of science has been utilized overwhelmingly to further the agenda of aggression and domination. Those who doubted the veracity of both science and religion, especially in their more ideological and mechanistic formulations, were essentially left out of the debate and shown to the cheap bleachers. Real life was in the hands of aggressive entrepreneurs, real men, captains of industry, with all their willing lieutenants in science, finance, and government. Professional evangelists (like Billy Graham) cozied up to such aggressive figures—consider Graham's relationship with Richard Nixon—thus establishing a highly visible connection between fundamentalism and secular power, a connection that served to justify civilization's "traumatic institutions," its Class and War "diseases," including its raw and brutal military aggression against "communist" countries such a Vietnam. The "Just War" doctrine.

But science, dominated by positivist reductionism, has been followed by its baggage, just as civilization's shadow has always been darkened by the "externalities" of utopia. As Paul Hawken says, "We live in a time in which every living system is in decline, and the rate of decline is accelerating as our economy grows." Pollution is the toxic shadow of progress, just as ecological degradation and various forms of slavery are the bloody footprints of civilization. Life and reality are not cleanly wrestled into utopian or "progressive" submission.

And so we find that better living through chemistry gives us grave problems with environmental poisoning; highway mobility and the internal combustion engine give us smog, mechanical death, rapid consumption of fossil fuels, global warming, and climate change; atomic energy gives us a glimpse into our own mutation or evaporation while building larger and larger pools of radioactive waste deadly for thousands of years; genetic engineering brings us to the tree Eve and Adam were denied in Eden—the Tree of Life. And so we confront in stark reality what we thought we had left behind: myth and its meaning in our lives.

III

The current political climate in the United States parallels, reflects, and is hugely shaped by the current religious climate. Liberals are tired and depressed, especially in the aftermath to September 11 as new programs for the pinching of civil liberties were passed by Congress with little dissent, as the wars in Afghanistan and Iraq were entered into with such wanton hubris, and as the financial meltdown of 2008, built on two relentless decades of magical capital growth, threatened (apparently) the American foundation of global capitalism. Aware of (though always hesitant to politically acknowledge) a long history of U.S. bullying abroad, liberals have become somewhat less eager to push buttons for progress, as conventionally understood. Conservatives, meanwhile, are increasingly pedantic, strident, and doctrinaire, throwing hell-fire ideological tantrums before the angry Father cracks down on all the wishy-washy liberals and their wicked social programs. Liberals are finding it increasingly difficult to believe, except as acts of stoic and robotic will, in the ideology of continually accelerating material progress. The great industrial machine has too many discontinuities and unanticipated problems—economic "externalities," for instance, chronic lack of equity in the distribution of wealth and policy determination, global poisoning—and, within the last decade, on American soil, acts of overt terrorism and financial meltdown.

Despite these problems, liberals remain unable (or is it just unwilling?) to deeply challenge the institutional status quo.[VI] Conservatives keep insisting that the machine of progress *will* continue to produce unimaginable abundance if only the freeloaders would get off their duffs and environmental Luddites would quit jamming the gears. Current political language is overwhelmingly shaped by a shallow but fierce contrasting of civilization versus terrorism, healthy capitalism versus diseased socialism, with all persons, all countries, told to choose which side they're on. That today's terrorist is yesterday's freedom fighter is a thought not welcomed in current discussion. In this respect, Reaganomics explicitly represented—and the Bush league form of Reaganomics continued to represent—an attempt to sustain the myth and the machine of progress under the moral umbrella of a religious conservatism that is often also fundamentalist.

Fundamentalist religion proclaims the world fallen; industrial and reductionist science proclaims the Earth inanimate. These worldviews—inherently masculine in their civilized assumptions and metaphysical assertions—coalesced tightly in the Reagan administration and have been continuous and essentially unchallenged since. So we had a James Watt who said openly that we may exploit as we wish for the End of the World is near, anyway—as if, at the end of time, the most spiritual thing we could do is strip-mine more coal. And we had eight years of a Clinton/Gore administration that could only cheerlead the "electronic superhighway into the twenty-first century" while simultaneously dismantling welfare,

---

VI. As I edit these essays prior to publication, it looks as if a recall election for Republican governor Scott Walker will occur in Wisconsin in June of 2012. Despite an initial upsurge against Walker's union-busting policies, it remains to be seen whether the Democratic Party will go beyond promising restoration of the status quo ante, address the realities of climate change and peak oil, and begin to grapple with what these unavoidable realities require.

overseeing an astonishing rise in the rate of incarceration (especially for minority males), and promoting policies (like NAFTA) that facilitated an enormous accumulation of new wealth for the already wealthy as it generated a huge increase in illegal immigration from the South, an emigration driven by economic desperation created by NAFTA. With George W. Bush it was tightened security, war, tax cuts, ecological indifference, and more deregulation. With Barack Obama it was profound disappointment for liberals who expected strong, humane change (for the most part) but who got (also for the most part) an uninterrupted continuation of empire, with gigantic financial bailouts for banks "too big to fail." (We might recognize in this an unexpressed conviction that civilization is also too big to fail—perhaps as dinosaurs were once too big to fail.)

Meanwhile, several seemingly independent and unrelated social movements have been groping toward synthesis. Among these movements are minority causes in various manifestations, the peace movement, the women's movement, the ecology movement, the same-sex movement, the continuous struggles of Native Americans for some sort of cultural reformulation, the alternative education movement (including home schooling), the commune, homesteading, and appropriate technology movements, the local-food movement, and, not least, a chastened socialism broadening its base (and its environmental comprehension) into Green coalitions. Neither the old-line liberalism nor the "new" conservatism commands much respect or holds much relevance for those who are sympathetic to or engaged in these creative social forces. People are, in their actions, moving out of and beyond the old myths, but a new myth has not congealed. A large and growing constituency is there for a new myth, but the new myth has not emerged. Emotionally, intellectually, and spiritually, people are milling around—waiting for a new dispensation but unable to congeal in a new unity.

There is, nevertheless, a growing intuitive consensus among those outside of the prevailing liberal/conservative conventions that ecology, the groundedness of indigenous cultures, feminism, racial equality, the renewal of rural culture, and the egalitarian essence of socialism are key elements in an emerging mythic synthesis. Yet people are properly wary of simplistic applications that produce a shoddy mythic collage.

The problem, or some of the problem, seems to lie in the area of religion or spirituality itself. That is, many leftists are agnostic or atheist, and religion often seems to them hardly more than escapist nonsense—the mass hallucination of ignorant, cowed and unenlightened people, the ultimate expression of unnecessary and stupid deference, the final denial of the earthy here-and-now. Religion, for them, is nothing more than a bogeyman form of thought control, political police garbed in "supernatural" robes, a great restraining force on the human imagination and therefore a restraint on political creativity. It therefore becomes nearly impossible in such a context to talk positively about spirituality; the subject is immediately suspect and instantly linked to religious beliefs or religious conceptions that not only fail to hold up to rational scrutiny but have already been rejected as irrelevant, misleading, or as social control devices of service to the ruling class or to male gender "superiority." Social conformity and political passiveness are seen as direct products or outcomes of religious deference. Insofar as churches (or denominations) are metaphysical salvation clubs, the charge against them—that they wallow in unethical deference—is a true one.

If one has uniformly rejected religion as prehumanist poetry (at best) or as authoritarian mind games in the service of class or sex domination (at worst), there then seems little room for consideration of any sort of mythic symbolism with cultural usefulness. Wanting to be trapped neither in positivist "single vision" nor in religious literalism, a great many people, especially ethical secularists, remain extremely wary of myth. To flirt with myth seems akin to dabbling with the occult—something powerless, escapist, quietist, and bizarre. This understanding sees the possibility of creative social change only in the realms of politics and popular culture. Spirituality is a total waste of time, a predictable dead end.

Yet myths carry an etiological purpose. If they are comprehensive enough, they reveal a symbolic assignment of causes and priorities. An emerging myth may present an alternative conception of social coherence. But how would a new myth come into being? Could it be free of "founding violence"? And how would it become absorbed in modern culture? We reject "supernatural" revelation; we are also wary of contrived "scenarios." The only option seems to be what Norman O. Brown, in his Introduction to *Life Against Death: The Psychoanalytical Meaning of History*, called for—a visionary synthesis of religion, psychoanalysis, anthropology, economics, and history. This visionary synthesis would have to be accessible to rational skepticism as a viable paradigm of cultural evolution; it would also have to be accessible to spiritual sensibility. That is, it would have to satisfy both rational skepticism and spiritual intuition. If we could successfully identify dominant trends and "archetypes" within the larger cultural history of civilization, then the creation or discovery of a new myth might be a surprisingly simple matter.

On the other hand, the prevailing Abrahamic mythologies seem incapable of surpassing the limitations and boundaries of their inherited mythic expressions even as they are woefully inadequate in addressing or incorporating into their mythologies the vastly enlarged historical and scientific knowledge at their disposal. In their most "conservative" forms, they are explicitly prevented from any such surpassing precisely because the religions themselves are deemed complete and sacred—their doctrinal articulations have been exquisitely honed in the direction of perfect or near perfect comprehension—and are not to be tampered with short of sacrilege. (Perhaps because religions have sharply distinctive "sacred" mythologies, their internal coherence must at all times be protected from sacrilegious tampering, both from within and from without. But civilization is not exactly based on "supernatural" figures whose Truth is absolute, but rather on sheer aristocratic arrogance acting with brazen willfulness and sociopathic élan. In that regard, all civilizations are as alike—"one another's contemporaries," as Arnold Toynbee put it—as religions are unalike. Religious zealots believe they're protecting God's Truth. Civilized elites simply assert that they are already in the gated community of the cosmic elect.)

This internal refusal to understand what's happening or even admit that it is happening is what we are now up against, globally, compounded enormously by the pervasive civilized schizophrenia made up of institutional predation (the class system operating through militarism and the commodification of nature) and utopian fantasy (a combination of cultural hierarchy, missionary ideology, and technological perfectionism). Whether religious ideology or civilized ideology is stronger is a hard question to answer. Osama bin Laden and George Bush accused each other, and each other's system, of entrenched and pervasive evil, and both were correct. Neither saw the log in his own eye. Such humble vision was not so much beyond them as it was beneath them.

A new myth may require neither an overarching conventional theism nor a stubborn "scientific" atheism as a precondition. A new myth may present a cultural model of human transformation (both personal and social, as myths are supposed to) that delineates a symbolic narrative of historical process. A myth needs to work also as insight; it must provide a broad and easily understood conception of the interconnectedness between the dynamics of social evolution, emerging human purpose, political mood, economic structure, ecological integrity, and cultural transformation. A myth is in some measure a mind game; but all myths are always also that. They are stories by which we humans explain to ourselves the immense realities around us and find our place in those immense realities. Mind game or not, myth is serious business. Myth is the Gestalt of cultural coherence. We can't do without it. We are, culturally, creatures of myth. It may not be farfetched to assert, along with the eclectic historian William Irwin Thompson, that *history* is myth.

We certainly need a new or reformulated myth that enables us to deflect or restrain our current speeding toward the global disasters now looming so unmistakably on the historical horizon. Spiritual transformation—and a new myth—have become political necessities. In our globalized predicament, there will be no adequate spiritual or political transformation without an accompanying and enabling myth; and there will be no new healing myth without the corresponding shrinkage of all three Abrahamic myths simultaneously with a fearlessly truthful reappraisal of what lies at the core of "civilized values." That is, to begin to deal adequately with the crises requires a simultaneous deepening in understanding of both religious myth and civilizational myth: not only recognizing any enchanting story of founding violence but, more importantly, recognizing what the founding violence sought to destroy, control, or repress. It's not merely the founding violence that's at issue here, but the nature and quality of the energy that the founding violence impounded.

IV

There is obviously an immense—and immensely explosive—chasm between religious and nonreligious people, with special regard toward those people for whom religion means fundamentalist belief. Each group—fundamentalist and nonfundamentalist—has come to see the other's worldview as either completely false or utterly incomprehensible. At the present time, there is little ground for useful dialogue. If anything, the chasm is becoming—if not increasingly wide—then at least increasingly rigid and brittle. And in a book like *God at 2000*, with essays by Karen Armstrong, Marcus Borg, Joan Chittister, Diana Eck, Lawrence Kushner, Seyyed Hossein Nasr and Desmond Tutu, it becomes quickly obvious that it is often easier to achieve dialogue between different religions (in this case Islam, Judaism, and Christianity) than between fundamentalists and nonfundamentalists within the same religious tradition. This is strikingly similar to what the Trappist monk Thomas Merton experienced when, in his *Faith and Violence*, he said that a Western Christian contemplative can "say he feels himself much closer to the Zen monks of ancient Japan than to the busy and impatient men of the West, of his own country, who think in terms of money, power, publicity, machines, business, political advantage, military strategy...."[1]

There are always a few highly respected social activists, like the late Dorothy Day, who focus on key ethical flaws in the larger society and who, by their moral tenacity in behalf of the poor and neglected, give to conventional religious institutions a better press than deserve. (Kathy Kelly, of Voices for Creative Nonviolence, has done this in behalf of the people of Iraq by calling attention to the deadly impact of U.S.-sponsored economic sanctions and ecological criminality in the aftermath of the first Gulf War. Now she and her colleagues are attempting to call attention to the disasters in Afghanistan.) These activists are, in a sense, the "saints" who redeem the pitiful ethical lethargy of those very religious institutions from which they emerged or to which, for whatever peculiarity of personal history or temperament, they find themselves attached.

Yet, to most honest nonreligious people, the powerfully committed actions of a few dedicated saints do not redeem the religious institutions. On the contrary, such actions serve to highlight the moral timidity and ethical vacuity of the institutions, as well as their subservience to the ethos of industrial affluence and deferential civility—mere property managers, as Phil Berrigan says of Catholic bishops.

Both the quest for salvation and the pondering of ethical considerations have become either privatized or etherealized; the legends and mythic tales of the past are held up either as literal history or as airy allegories. For fundamentalists, one's private belief in these literal accounts safeguards one's salvation. This mix of internal privatization and external literalizing seems to be characteristic of all fundamentalisms. The "faithfulness" with which these beliefs are clung to is fuelled largely by a fear the examination of which is prevented by the strictures of the religious ideology itself. To question belief is tantamount to dabbling with the Devil. Therefore to break free of fundamentalist conviction is a terrifically painful and risky process that puts one's very soul in jeopardy: fear of God's wrath, fear of the Devil, fear of eternal hellfire.

Yet for most nonreligious people, such events as Moses receiving the Ten Commandments directly from Yahweh on Mount Sinai, the supernatural conception and subsequent physical resurrection of Jesus, and the direct selection by Allah of Mohammed as the final prophet of true religion, are all historic fictions. That is, those events did not literally occur. *Something* may have happened—powerful events that generated compelling stories—but whatever really happened is shrouded in mystery and buried under legend. Nor does whatever happened merit the cloak of divine finality associated with it by its fundamentalist and metaphysical partisans.

For a great many religious people, however, these events—or some specific selection of them—*did* literally and actually occur. Furthermore, an accurate grasp of *all* historical meaning is contingent upon the recognition and acceptance of some pivotal "historic" events peculiar to each religious tradition: totally true in one's own tradition, totally misunderstood if not totally false in the others'. No matter how stupid, absurd, stubborn, or wicked each position may seem to the opposing camp, there is no denying the political power and explosive divisiveness of the contending viewpoints. Insofar as each of the Abrahamic religions constitutes not only a distinct cultural development but also a geopolitical formation, the stage is set for horrific escalations in conflict and violence: real grievances built on metaphysical fault lines enormously compounded by contending religious loyalties, ideological hypocrisy (including the masking of capitalism as "freedom"), and unbridled, righteous male energy in possession of apocalyptic weaponry.

What's at stake to the secularists is the growing threat of powerful religious fascism, a seizing of political control by religious ideologues in order to impose the will of God. What's at stake to the fundamentalists is the pervasive fear of the Father's condemnation, due to a "godless" secularism allowed to run unchecked. One looks in vain, it seems, for a point of contact. This is a frustrating puzzle, for we are, by and large, descendants of a common mythology, or at least descendants of mythologies with a common patrimony. It follows from this that we would do well to look for a ground to talk upon not so much in the specifics of belief as in the generality of myth. Intellectually and spiritually, there is a membrane between and around the spheres of belief and nonbelief that only myth, partaking literally of neither belief nor nonbelief, can enter. Myth is, on one level, simply a story descriptive of cultural origins and social purpose; and, as such, myth simply *is*. Therefore only a new myth or new cultural story that both fulfills (though unexpectedly) the old myth and simultaneously inspires and justifies a new pattern of human conduct can hope to bridge, or perhaps even resolve, the apparently ineluctable confrontations. We need a larger, more all-embracing myth. We need a new cultural story. And we need it very soon.

V

The rise of the conservative Right in America is obviously a very important political and religious phenomenon. This "new" Right, in the policies of the Republican Party (as in the Clinton-era policies of "new" Democrats), embodies the essence of a masculine ethos that apparently sees itself defending and promoting conventional capitalist prerogatives. In the "new economy" we won't need much of a "safety net" because the economy itself, as Newt Gingrich has told us, will be a dynamic trampoline. The capitalist economy, like religious evangelism, is energetic; each demands room for its missionary expansion. Each wishes to convert individuals, even entire nations, to its respective worldview; but neither is genuinely democratic.[VII] Corporations, like churches, are far more feudalistic than they are democratic. Both ideologies—the capitalist and the religious—assert entrenched justification for their advocacies, policies, and actions. Both claim historical and religious precedent. Both assert the basic selfishness of human nature—although capitalism claims to usefully channel selfishness, a channeling with which fundamentalism apparently agrees.

It is, of course, ironic that "sin" in conservative religious ideology miraculously becomes "freedom" in secular capitalist thought. This irony is neatly unfolded by Fred Block, in an article entitled "The Right's Moral Trouble" in the September 30, 2002, issue of *The Nation*. Block describes how a "powerful alliance of economic and religious conservatives has set the political agenda":

> The political backstory to the rise of the Republican Right over the past twenty-five years is familiar. American conservatives were in the political wilderness through much of the 1950s and 1960s; deep factional divisions made it difficult for economic and religious conservatives to make common cause. But starting in the 1970s, conservative intellectuals worked to fashion a political ideology that would allow these divergent groups to coalesce under a single umbrella....
>
> In shaping this ideology, these intellectuals had to overcome the deep distrust between religious conservatives and

---

VII. See Naomi Klein's *The Shock Doctrine: The Rise of Disaster Capitalism* for an unflinching look at traumatic institutions operating with world-changing energy.

> economic conservatives.... Religious conservatives were deeply suspicious of the materialistic pursuit of wealth.... Moreover, some of the most important theorists of "free markets" were overtly hostile toward organized religion because of its insistence that there was something "higher" than the pursuit of material self-interest.
>
> The trick that intellectuals used to reconcile these conflicting viewpoints was to treat "the Market" as something like a divine force that always calls forth moral behavior. They took Adam Smith's metaphor of the "invisible hand" and used it in an audacious way. "The market" became a natural force to which people had to accommodate.... And liberalism and socialism were obviously in error because they were godless efforts to interfere with the natural and divine logic of the market.
>
> This was a brilliant formulation because both sides could read into the argument exactly what they wanted. Religious conservatives could imagine that divine providence, working through market institutions, would compel individuals to follow the Ten Commandments. Economic conservatives could use this rhetoric to delegitimize government regulation of business and push for lower taxes.[2]

What an irony that capitalism, in its early industrial assurance, found its moral base in the secular ideology of Social Darwinism, and that socialism rests, ethically, on the great principle of sharing! But fundamentalists are not particularly noted for the vigor or subtlety of their intellectual probings, and they therefore simply ignore or are just ignorant of their ideological inconsistencies. (And now hiding these inconsistencies behind fierce denial has become a political necessity.)

Fundamentalism claims direct inspiration from the Ruler of the Universe as conveyed through a volatile mixture of apocalyptic symbolism and flat-footed literalism as "taught" by the Torah, the Bible, and the Koran. In the same issue of *The Nation* in which Fred Block's article appeared, there is also an essay, "Fascism's Firm Footprint in India," by Arundhati Roy. Ms. Roy deftly describes the rise of a kind of Hindu fundamentalism whose energy is both fuelled and directed against the minority Muslim community by right-wing political leaders:

> Historically, fascist movements have been fueled by feelings of national disillusionment. Fascism has come to India after the dreams that fueled the freedom struggle have been frittered away like so much loose change. Independence itself came to us as what Gandhi famously called a "wooden loaf"—a notional

freedom tainted by the blood of the hundreds of thousands who died during Partition. For more than half a century now, that heritage of hatred and mutual distrust has been exacerbated, toyed with and never allowed to heal by politicians. Over the past fifty years ordinary citizens' modest hopes for lives of dignity, security and relief from abject poverty have been systematically snuffed out. Every "democratic" institution in this country has shown itself to be unaccountable, inaccessible to the ordinary citizen and either unwilling or incapable of acting in the interests of genuine social justice. And now corporate globalization is being relentlessly and arbitrarily imposed on India, ripping it apart culturally and economically.

There is real grievance here. The fascists didn't create it. But they have seized upon it, upturned it and forged from it a hideous bogus sense of pride. They have mobilized human beings using the lowest common denominator—religion. People who have lost control over their lives, people who have been uprooted from their homes and communities, who have lost their culture and their language, are being made to feel proud of something. Not something they have strived for and achieved, but something they just happen to be. Or, more accurately, something they happen not to be.[3]

The old classical aristocracy used raw force to impose its will. But the new commercial aristocracy, now that civilization has defeated or destroyed all noncivilized forms of economics and marginalized all forms of spirituality inconsistent with its utopian worldview, operates in a political atmosphere requiring "democratic" support—at least so long as "democracy" remains a cultural shibboleth. The new "conservative" aristocracy finds and stimulates support by tapping into broadly held religious mythology. Political conservatism leans heavily on religious conservatism for maintenance and justification. This seems to be at the heart of the Constantinian Arrangement. This entire psychopolitical complex is tied into a structural matrix of masculine prerogative. Religious institutions are both lightning rod and moral whip for the omnipotent Father God. Society as a whole is kept in line beneath threatening images of Father wrath. The prison "industry" controls racial rebellion—rebellion from below. Men assert dominion over nature and maintain a public ascendancy over women, all of it "justified" by the divine Maleness of God. Class stratification derives heavily from this mythological inheritance.

With a view of all people as inherently sinful and (except for born-again Christians) doomed to perdition, Christian fundamentalists of necessity maintain a pessimistic understanding of the ultimate value of democratic government or human self-regulation. Only firm control on the part of God, through the agency of God's legitimate representatives, can maintain a semblance of political morality and an ordering of social decency; only religious salvation is potent enough to forestall personal and perhaps even cosmic disaster. Democracy is only as good as the Christians who hold office and power. John Nichols, in a sidebar article called "Dark Ages Ahead at the NLRB," in the September 3/11, 2001, issue of *The Nation*, writes of the Christian fundamentalist organization American Vision. Nichols says "The Georgia-based group's president, Gary DeMar, preaches about 'the necessity of storming the gates of hell' and cleansing public institutions of 'secularism, atheism, humanism, and just plain anti-Christian sentiment.'"[4] Only final authority from the throne of heaven, in other words, can hold the line against the drift of natural instincts, human depravity, permissiveness in immoral society, and the sneaky intransigence of fallen nature. Such authority requires strong male control; it requires an authoritative Father, whether in the family, the nation, or the pantheon. The righteous Father and the obedient (but also righteous) sons of the Father must discipline the unruly and punish the wicked.

Religiously this Father authority requires worshipping the Father in heaven and exalting the father in the family. Politically it entails a powerful police system to crack down on immoral behavior (the "War on Drugs") and violent political opposition (the "War against Terrorism"). Economically it justifies continual exploitation of natural resources, the holding down of wages, and the shredding of social programs. Intellectually it means the determined expansion of a globalizing capitalism that is always growing. All this is empowered and protected emotionally by a diffuse mythic tradition of male prerogative; and this tradition holds sway despite inconsistencies, contradictions, and the looming threat of global catastrophe. (In fact, looming catastrophe is sometimes utilized to demonstrate the prophetic power, and thus the inherent wisdom, of the male mythic tradition. But catastrophe is *never* the fault of the righteous.) The basic thrust of fundamentalism in political life, therefore, is to use the *demos* to reinstate, in some variable particulars, the *deus*. The *deus* or divinity to be reinstated is, of course, the King of Heaven. To reinstate the rule of heaven is to re-establish the Fatherhood of God as a political fact—one nation under God.

It's not incidental that Thomas Jefferson, probably the foremost exponent of participatory democracy among the founding fathers, was a Deist.

The Kingdom of God Is Green

That is, Jefferson believed in a doctrine placing God outside the realm of direct intervention in human affairs. In his Introduction to *The Gospel According to Jesus*, Stephen Mitchell says "Jefferson, too, called himself a Christian. 'To the corruption of Christianity,' he wrote, 'I am, indeed, opposed; but not to the genuine precepts of Jesus himself.'"[5] God for Jefferson was, poetically, "Nature's God." Yet Jefferson was also fascinated by the orrery, a fashionable contraption of enlightened high society depicting the universe-as-machine. This fascination led more directly to a conception of God as Master Mechanic or Master Architect than, as Nature's God, to Gaia as sentient Earth Mother—or simply as Spirit, whose "nature" we can only begin to touch via true humility and deep reverence. (Deism, it seems to me, keeps God in a distant shop, tinkering somewhat absent-mindedly with a vast, complex machine called the Universe, while theism depicts God as a supernatural and supercivilized vigilante constantly on the prowl in behalf of a pagan eradication program.)

For all the lofty abstractions about God, only a scattering of "mystics" and Anabaptist communities have ever taught that the nonviolent and noncoercive "kingdom" of God, as propounded by Jesus, was descriptive of how people could and should actually live. This teaching, this practice, has been, and still is, too radically simple, too ethically demanding, for either fundamentalists or religious liberals to accept. Both fundamentalism and mainstream religious liberalism are, in fact, largely *evasions* of the kingdom of God. By making an idolatry of "right belief" and biblical inerrancy, fundamentalists avoid the existential demands of kingdom behavior. By making an idolatry of "civilized values" and "progress," religious liberals do a parallel thing. Thus the self-imposed limitations of conventional Christianity (both liberal and fundamentalist) serve to degrade, diminish, and discredit the teachings of Jesus, especially and particularly in regard to the meaning and ethical unfolding of ecological community. Religion therefore becomes an obstacle, an impediment, to the ethically committed. Without a grounded vision of the kingdom of God within the church, the more ethically committed drift into ethically based secular organizations, even if the basis for secular ethics remains weak in its intellectual foundations. (Not that the ethical impulse is wrong—far from it—only that "because it's right" or any other back-to-the-wall assertion of justification lacks concrete historical coherence and may even lead to disaster, as with Soviet communism.)

E. P. Thompson, in *The Making of the English Working Class*, shows repeatedly how the proponents of broader democratic government in

England, toward the end of the eighteenth century and the beginning of the nineteenth, were Jacobins, Deists, and atheists. Correspondingly, Thompson shows how the growth of Christian revivalism, with its keen emphasis on personal salvation in an afterlife of bliss, was a drain on popular democratic militancy. We easily forget that those groups and individuals who won what democratic advances we have—including the Bill of Rights—were often entrenched opponents both of organized religion and of conventional religious intervention in political life. And it remains true that much of the popular hesitancy regarding democratic principles and democratic action—from practical personal liberty to structural economic equality—is grounded in psychocultural deference or reflexive yielding to the moral strictures of the Judeo-Christian Father. What we see here is the astonishing paradox that much of the ethical thrust of the Gospels is advocated most forcefully by agnostics and atheists and denounced most stridently by "believing" Christians. That is, because the "kingdom of God," mentioned scores of times in the first three Gospels, has been largely deleted from conventional Christian discourse, its implicit ethical energy continues on in a largely intuitive but inarticulate form by agnostics and atheists, even as "believing" Christians are so deeply poisoned by religious doctrine (e.g., "original sin") that they reject and scorn the very transformative energy contained within and conveyed by the "kingdom of God" proclamation. Needless to say, "believing" in a set of doctrinal statements about ultimate reality does not necessarily correlate to an unfolded inner communal life based on stewardship and sharing.

Religious fundamentalists can, in the end, only view democratic procedures as provisional tools in the struggle of believers to re-establish the forceful authority of the *deus* and of religious (or religiously blessed) authority in general. It's true that religious dissenters and nonconformists were heavily involved in the effort to dismantle, or at least redirect, the centralized, elite control of both church and state. Present-day fundamentalists, no matter what the particulars of their religious persuasion, can be construed only tentatively as committed participants in a genuine democratic process in pluralistic society, or as upholders of the humanistic values that give democracy much of its essential meaning. No matter how splintered fundamentalist groupings may be (often over contentious points of doctrine or "prophecy" that others simply find bizarre), the magnet that polarizes the entire field is the powerful conviction that an authoritarian male God simply must be obeyed.

The Kingdom of God Is Green

Fundamentalism—whether ethnic and "racial" in Hitler's case, political and economic in Stalin's, or religious and cultural in the case of the Ayatollah Khomeini—can move with bland self-assurance toward overtly fascist forms of control. Christian fundamentalism does not differ greatly in its ultimate political purpose from Islamic or Jewish fundamentalism: each seeks to re-establish the rule and righteousness of the Father God over society as a whole—and in self-consciously civilized societies, these efforts are therefore carried out through civilized institutions. These fundamentalisms contain concentrated ideological opposition to secular freedom and gender equality.[VIII] That the "kingdom of God" might be a kind of social yeast growing from *within* secular (but intensely ethical) movements is simply beyond the ken of fundamentalists. But it is also critical to recognize that "civilized values," even though women and minorities have achieved a significant degree of legal equality within the larger civilized system in recent decades, have at their core a structure of male control that is both military and economic. This is what Arnold Toynbee meant by Class and War being at the heart of civilization, and what Lewis Mumford meant by "traumatic institutions" being at the core of the civilized inheritance.

It is important to note that what is identified as sinister by religious fundamentalism is closely associated with, or seems to fundamentalists to be closely associated with, either the excesses of political liberty or regression into "paganism." Pornography, for instance, is seen to be the result of a permissive, degenerate, and irreligious society lacking in self-control. It also appears to be the case that erotic allure is projected onto women by men, so that the resulting lustful feelings men feel (and fiercely suppress in themselves and in each other) are identified as being caused by female allure. Thus is "ownership" or responsibility deflected and denied. Humanists respond by asserting that pornography, prostitution, rape, wife battery, and child abuse are often the tormented products of cruelly compressed patterns of sexual denial and body hatred whose roots are firmly entrenched in the religious and economic organization of a stern and antifeminine patriarchal domination—a domination in which religious

---

VIII. Christian Parenti, in an article called "Ideology or Electricity" in the May 7, 2012, issue of *The Nation*, page 32, says that in Afghanistan the "so-called Red Prince, Amanullah Khan, who ejected the British in 1919, was dethroned ten years later by a tribal rebellion that opposed his Turkish-inspired modernization efforts. He had imposed a modicum of land reform, given women the vote and started educating girls. Rural elites would accept good roads, but not the taxes to pay for them; the rural masses would accept agricultural improvements and education, but not an assault on patriarchy."

fundamentalism itself represents the most concentrated psychological expression: man in God-given mastery over fallen nature.[IX] Repression generates both an obsession with order and control as well as setting the stage for its opposite, which is disruption and chaos. But since capitalism, for instance, has divine protection, as Fred Block has shown, fundamentalists are incapable of tracing society's brokenness to the rise, spread, and cultural penetration of capitalist economics; they are ideologically blind to the very process whereby the market system has corrupted and destroyed countless forms of cultural stability, while repression characterizes the necessary order demanded by God.

From inside their self-contained morality, sanctioned by long tradition, fundamentalists perceive their own values and character structures not only as normative but confirmed both by the Bible and by God. Nothing could be clearer or more absolute. Any challenge to this normative posture gets shrugged off as an attempt by Satan, the Devil, to test, undermine, subvert, or otherwise corrode divinely ordained patterns of proper conduct. Condemnation by the Father (and, therefore, condemnation by the social peerage of all religious fathers and brothers) is to be avoided at all costs. Fear of personal condemnation and the perceived threat of identity loss underlie the psychological power of all forms of fundamentalism. Going to hell forever is the ultimate disgrace and fear.

There is therefore also a religious component, a mythic component, in the contraction of capitalist civilization. It is not only the capitalist dollar and economic prerogative of the West that are in process of being squeezed, but the religious hegemony of the Father God as well. What is psychologically and culturally at stake are the historically revolutionary questions of gender imbalance, racial discrimination, class domination, economic rapaciousness, and ecological degradation. To more deeply democratize society and develop global economic ecology is not merely a political revolution; it is also a cultural and religious revolution. But for this we have, as yet, no myth. All we have, on the one hand, are contending versions of a more or less common myth whose enraged partisans drive us toward catastrophe and, on the other hand, a broad-based network of social change advocates who lack political traction in large part because they are devoid of a unifying myth, a compelling story. The former represent concentrated and historically accrued male power, fiercely unwilling to relinquish that power; the latter are representative of a cultural ethos

---

IX. See also the Catholic critiques of Garry Wills in *Papal Sin: Structures of Deceit* and Peter De Rosa in *Vicars of Christ: The Dark Side of the Papacy*.

whose feminine values are bubbling to the surface, as in a yeasty bread sponge not yet congealed as dough.

The fear of sexual equality (which is only in part a fear of greater erotic expressiveness), in concert with the continued protection of economic, class, and racial advantage, has already resulted in a fabulous entrenchment of social conservatism supported by biblical literalism—witness the stubborn contempt of Bush junior and the embarrassing weakness (or outright deceptive misrepresentations) of Barack Obama. The wholeness of quiet faith in the abundance and essential goodness of natural life is being trampled under the fearful, insistent rush to build an impregnable fortress in defense of those beliefs that cast nature as fallen and human beings as wicked. This cultural armoring expresses itself economically through the extension of the industrial machine built to the specifications of reductionist science. Star Wars or the Missile Defense Shield represents this impulse driven to absurd theatrical parody—as if the United States could be magically protected from an atomic Armageddon that would engulf the rest of the world. The War against Terrorism becomes both a Christian crusade against false belief and a civilized crusade against backwardness.

There is, obviously, little common ground for serious dialogue between humanists and fundamentalists. Yet it may be possible, however tentatively, to find such a ground. What's needed is a myth, based on real evidence but not narrowly positivist, offering a fresh and alternative conception of historic purpose that, in some measure, can fulfill the old myth—not by destroying that myth, exactly, but by filling in its voids.

Let's proceed to a stranger analysis.

VI

I have the framework of a myth to propose. The outlines of this myth congealed over a period of years through the combined processes of intuitive insight and rational understanding. The key discovery, oddly enough, was a direct result of my determination to understand the origins of agriculture and why its fate, at least since the emergence of civilization, has been so closely associated with slavery and other forms of bondage or inferiority. I do not ask anyone to "believe" this myth as if it were empirically factual, as if Spirit actually wears—or is—these masks. I do ask that its implications be given serious contemplation and quiet consideration. Its outlines conform to real history. There is no precondition of conventional, extraterritorial supernaturalism to accept. Its primary requirement is not

Age of the Daughter

the swallowing of a newfangled idea but the acceptance of an ancient (but denied and ignored) historic fact.[X]

Let me go on being personal and confessional for a moment. I feel rather uneasy proposing a cultural myth. Nothing could seem more ridiculous or absurd than an individual propounding or promoting a new myth. Myths are not private inventions. Yet I did not make this up out of nothing; the elements were all around me (as they are around us all, buried in history and visible in the everyday mundane), and they fit together so simply and so well that I found myself unable to brush them aside as irrelevant and unrelated. I *discovered* these elements; I did not *invent* them.

We have begun to talk now of wholism, of inter-relatedness, of ecological systems, of Gestalt. Ideas of circles and phases and mandalas now contend for our attention along with straight lines and organizational charts and triangles. In architecture we have begun to move from sheer cerebral design to what Christopher Alexander and his architectural colleagues call "pattern language." We are rediscovering organic agriculture, small-scale community, do-it-yourself simple living. As utopia more fully reveals its demented energy and destructive sterility, we begin to yearn for earthier and more humble eutopian alternatives. The vastness and complexity of evolved life begin to sink in on us as we hear, read, and think about species extinctions, the potential end of higher life forms on Earth if the thermonuclear arsenal is unleashed—or even, like the polar bear, extinction due to global warming. (There is the mere beginning of a critical discussion on climate change as a consequence of arrogant utopian economics.) The straight line of progress, the triangle leading fiercely upward from the uncivilized and the crude toward pure urbane civility, now appear increasingly mocking and destructive. Utopia, we find, is toxic. We discover that we begin to think again in curves and whorls and circles. Faced with the potential end of our linear civilization, we begin to explore, even unconsciously, the possibility of eternal recurrence or the ecological unfolding of eutopia. And, in a totally new way in America, we are now terrified of terrorism and increasingly desperate for a worldview—a myth—that makes sense of this global disaster and points a way out.[XI]

X. If Israel's identity is based on a myth of "founding violence"—Hebrew genocide perpetrated against the residents of Jericho—civilization's "founding violence" is the armed robbery of the agrarian village. Without confession and repentance, the consequences of such foundational acts continue on and on and on. "Christian civilization" merges these two myths in world-conquering mission.

XI. Although this is impossible to verify empirically, it seems to me that the current upsurge in the fear of terrorism may not be significantly different, emotionally

The Kingdom of God Is Green

We are, in a sense, growing out of old myths because there is no alternative but to grow out of them. We are being *forced out* by history. But coming out of the grid, we rediscover the eolith. Tired of road rage and the superhighway, we are enchanted with the footpath. Sick of jet skis, we remember the kayak and canoe. Repulsed by the snowmobile, we dig out the snowshoes or the skis. What was old-fashioned and outmoded now has fresh charm—appropriate technology and (Carl Jung's term) reform by select retrogression.[XII] We begin to realize that to be masters of nature is to seal our doom. We now wish to humbly reconnect with both nature and the past, but on a higher cultural level and on a finer spiritual plane.

We are not immortal gods and goddesses. We are humble creatures who eat and defecate, female and male, who emerge from and return to the Earth as surely as a tree, a flower, a fish, a spider, a turtle, or a bear. We have big brains. We walk on our hind legs. We have clothes and tools and fire. But we have advanced, technologically, so fast in the last couple of centuries that we who are now alive face, literally, the extinction or transformation of our species. Therefore I come as humbly as I am able, wanting to be critically honest yet generous to all. If there is anything that others can glean from this "myth," then my own consciousness will have provided something more than an exercise in the randomness of futility.

VII

Despite my having stepped out of and beyond the formal religious indoctrination of my Calvinist and Lutheran upbringing, I carried into adulthood a great deal of puzzlement regarding the Holy Ghost. It had been fairly easy, in terms of the Christian Trinity, to imagine the Father or the Son—especially the latter, for his sentimental yet somehow stern portraiture was posted everywhere except the post office, the only public place a picture of Jesus would have hung in his own lifetime. But there was never so much as a hint as to the identity of the Holy Ghost, other than oblique reference to some sort of holy glow emanating from the intense reciprocal

---

speaking, than white fear of Indian "savagery" in earlier centuries. Cultural integration with the "terrorist" is not deemed acceptable or possible because of the terrorist's inflexible and degenerate wickedness. Thus we are relieved of the burden of agonized introspection or open-ended dialogue. Only in our time, crises have become epidemic, and so it's no longer possible to evade the consequences of our various imperialisms—except, of course, in the short term, by righteous denial.

XII. See my use of Jung's construct in "Organic Intelligence," Chapter 1, pages 7 and 8, in *Nature's Unruly Mob*.

love between Father and Son. But that definition never seemed adequate. Father and Son were Persons and Beings. How was it that the Holy Ghost had achieved that status? How does a Ghost or a Spirit get to be a Person? I had a feeling the Holy Ghost was in some sort of mythic closet, and nobody was daring—or was permitted—to peek in. The Holy Ghost made people decidedly uneasy. Pentecostals, supposedly under the influence of the Holy Ghost, handled snakes and talked in gibberish; but this behavior only intensified conventional unease in regard to the Holy Ghost. The less said the better, apparently.

I became aware of the Berrigan brothers, of Dorothy Day and the Catholic Workers. I associated with radical ministers, nuns, and priests. I read some of the writings of Dietrich Bonhoeffer, the Lutheran clergyman whom the Nazis executed for his fringe role in a plot to kill Hitler. None of this helped very much, at least not directly, for the great bulk of these folks, whose ethical convictions I enormously admired, did not, for the most part, push beyond theological conventions, nor (and this is crucial) did they seem particularly interested in or troubled by the theological implications of prehistory. Before the rise of civilization, God appears to have been exceedingly bored. Prehistory, insofar as it was considered important or accepted at all, was just a set of museum pieces, mute and dusty. It was, *and it remains*, not integrated into Christian mythology.[XIII]

It was a process of sheer intuition—though I could as easily call it prayer—that led me to realize the Holy Ghost was feminine. I understood then why the Holy Ghost was kept in a closet: if God was Three Persons, and if the Godhead was totally masculine, then the Holy Ghost as feminine Person demanded the gender bending of God. Yet the orthodox tradition was clear. As C. S. Lewis had his hero Dr. Ransom say in *That Hideous Strength*: "What is above and beyond all things is so masculine that we are all feminine in relation to it."[6] *That* was the essence of the orthodox gender tradition, a tradition permeating and shaping the cultural and religious sensibility of all three Abrahamic myths. The Holy Ghost was secretly female, and the patriarchal officialdom of Christian orthodoxy was fearful to the death of—Her.

XIII. There is an interesting school of thought among ecologically oriented Catholics (we might say Thomas Berry is their patron saint), a school that has developed a cosmic mythology called the "New Universe Story." I will deal rather extensively with their ideas in a forthcoming work entitled *Picking Fights with the Gods*. For now I will only say that the New Universe Story is only incidentally about ancient, precivilized human history. Its real fascination seems to be with the origins and expansion of the universe. As such, it too fails to integrate precivilized spirituality into Christian mythology.

The Kingdom of God Is Green

But surely there is more substantial evidence for this assertion than idiosyncratic intuition? Surely there is.

VIII

In the twelfth century, the Italian monk Joachim of Floris described what he considered the three ages of human history, a scheme of "ages" he arrived at via a conceptual meshing of the Old and New Testaments of the Bible with the Christian Trinity. First, Joachim said, there was the Age of the Father. That age corresponded to Judaism, the Torah, and the Old Testament, and it was marked by monarchy, discipline, and law. Second was the Age of the Son, a period stemming from the life and death of Jesus, manifesting itself socially as love institutionalized in the church. Third, an era yet to come, there would be an age of holy freedom or consecrated anarchy, the Age of the Holy Ghost. (One can find commentaries on Joachim's influence in, among other works, the opening section of Michael Harrington's *Socialism* and in Karl Lowith's *Meaning in History*.)

Joachim's conceptualization of historical epochs is a valuable and useful mythic tool, even though incomplete. He gave metaphysical abstraction an earthly grounding. But Joachim knew nothing of what we now call prehistory. I believe we can say he thought the Genesis account essentially factual. The medieval church may have had threads and currents of allegorical understanding, but it would be anachronistic (if not downright deceptive) to suggest that allegorical understanding was widely or deeply applied to the Genesis account of Creation. Behind the orthodox understanding stands God the Father, Creator of Heaven and Earth. That is, in the orthodox understanding, the nature and character of God is accurately portrayed in Genesis, even if details regarding the creation story might be understood as allegorical. In our Western tradition it is indeed possible to identify the monotheistic religions of Judaism, Christianity, and Islam as the repositories of patriarchal theological prerogative they were and still are: social systems whose mythic core is a jealous and powerful Father, deference to whom, appeals to whom, are earnestly engaged in by both Jews and Palestinians, for example, in the seemingly endless conflict in Palestine and Israel, not to mention the larger and more deadly dynamic of Jewish-Islamic hostility at the level of contending nations. These patriarchal religions arose in a world in which male-dominated civilizations were the norm, where maleness determined power, where God had to be

Age of the Daughter

Male. The patriarchal religions "divinize" a male psychology powerfully dominant.

If the era of the Son was love institutionalized in the Christian church, the eventual secularization of this ethos produced the welfare state: love institutionalized in massive governmental bureaucracy—from, we might say, the City of God to the City of Man. We might also say that Gospel or spiritual energy—ethically—over time—helped create a more compassionate use of governance; but the forms through which compassion was expressed were largely shaped not by any eutopian cultural wholeness but by derivation from the existing utopian governmental template. Compassion worked to reform utopian constructs; such reform, however, was largely unfamiliar with any "kingdom of God" eutopian possibility. Furthermore, both Judaism and Islam trace their religious ancestry to the patriarch Abraham. The Jews do so through Abraham, Sarah, and Isaac; the Moslems do so through Abraham, Hagar, and Ishmael. (See Genesis 16–21.) Meanwhile, the Christian church provides a common heritage to both Western capitalism and Eastern communism through—what irony!—that converted (Lutheran) Jewish intellectual, Karl Marx. (Marx is to capitalism in a sense what Luther is to Catholicism.)

Despite the considerable period of time (roughly two thousand years) separating the respective emergence of Judaism and Islam, it's possible to say, in terms of their respective orientation toward the Father, that these two religions are *psychologically* contemporaneous. That is, they represent rival Father systems in which deference to the authority of the Father is the boundary of the permissible and the possible. This mythic deference, in addition to the actual male-control core of civilized institutions, is the true basis of what is loosely called male domination: actual political control of taxes, swords, and spears, all mythically protected by religious doctrine and imagery for use in worship of a male God.

The institutionalized love of the church also found itself, eventually, with two major economic expressions: capitalism and communism—private salvation versus collective salvation—both forms of secular salvation, however, shaped overwhelmingly by utopian science and civilized economics. The Christian church, from its decentralized "primitive" period after the death of Jesus, spreading throughout the Mediterranean basin, steadily abandoning the radical "kingdom of God" vision in favor of creedal assertion and metaphysical abstraction, became first absorbed by the Roman Empire and then its leading expression. When Rome fell, the church remained the leading instrument for religious expansion and civilizational unification. This is what theologian William Stringfellow has called the "Constantinian Arrangement." By merging with empire, the church completed its transformation from peasant parable to civilized ideology by rationalizing its accommodation to power. Theocratic power trumped—shall we say trampled?—the nonviolent transformative energy of the kingdom of God.

The Protestant Reformation announced a limited democratization of the church. The "brotherhood of man" now became both a religious and a political ideal: the "priesthood of all believers," provided the believers were male and held the right belief. As secular government took over areas of control formerly held by the church, the impulse toward institutionalized love grew into secular ideology. The dominant ideology was capitalism, upwardly striving, morally superior, and inherently elitist. But economic historians learned to see in the working-class struggles of the general Reformation period (especially peasant revolts and Anabaptist refusals) an inchoate striving toward a more comprehensive communal ethos, an awakening into social reality of Jesus' teachings regarding the kingdom of God. (We might even say that Reformation-era Anabaptists rediscovered the kingdom of God as the core of Jesus' teaching, but they did so with what amounts to a conventional, orthodox, and fundamentalist understanding of God in the mode of biblical literalism. That is, there was a true recognition of the significance of the kingdom of God proclamation, but the theology of that recognition did not break through from utopian to eutopian.) But industrialism and the concomitant destruction of agrarian society generated two self-conscious, rival economic ideologies: capitalism and all forms of industrial collectivism. Both developments can claim something of a Christian heritage as offspring of utopian Christendom. Capitalism is more directly theocratic and Augustinian. Collectivism has largely assumed these civilizational theologies, too; but it has added—in various degrees of commitment—the ethical thrust of sharing and caring. They are, therefore, secular manifestations of the Age of the Son, of love institutionalized. If the Right draws the line at charity, the Left goes on to advocate for redistribution.

Yet in the nineteenth century, beginning with the scholarly archaeological work of Johann Jakob Bachofen—whose studies greatly influenced Engels and Marx—the early roots of civilization began to be unearthed—or, perhaps we would more accurately say, the precivilized world of agrarian villages began to be unearthed. The data were startling. Underlying patriarchal civilization was a broad and deep stratum of life reaching from

Age of the Daughter

the Neolithic agrarian villages, a stratum that could only be called the Age of the Mother. The divinities of this period, from the Paleolithic to the fringes of the patriarchal, ranged from rough representations of female fertility figures in stone and clay all the way to theological conceptions of the Queen of Heaven. This is the period of history now under such intense study—and intense dispute—by feminist scholars. Whatever the ultimate conclusion about the quality of life during this stage of village agriculture, it remains true that the Age of the Mother was all but obliterated by the patriarchal religions that followed, as male control congealed and consolidated in the enormously powerful economic and military predations of civilization—in what we might, following Gil Bailie, call civilization's founding violence.

Anthropologists and historians of antiquity are, apparently, in general agreement that a great many of the cultural advances from the Stone Age onward toward the rise of the city can be attributed to women. These developments include agriculture (horticulture in particular), pottery, medicine, spinning and weaving, and—not least—the stability of settled village life. The stable village of the late Neolithic period, rooted in horticultural abundance, was the archaic economic core that was expropriated by the aristocratic city, which grew large and mighty on the "surplus" food production stolen from the village.[XIV]

So to Joachim's trio of ages we may add a fourth age—or, from a chronological perspective, really a first age—verified by modern scholarship: the Age of the Mother.

The increasing size and complexity of cities, the specializations of labor, the emergence of institutionalized warfare and explicit mass slavery, the invention of the plow, the development of wheeled vehicles and sailing ships, and the involuntary servitude of the coerced agrarian village as a whole, combined to press the social influence and communal prestige of

---

XIV. The Cain and Abel story (Genesis 4:1–16) depicts the shepherd (Abel) murderously victimized by the farmer (Cain). This is—and has generated—clever historical revisionism. That's not to say that there wasn't conflict between nomadic herders and settled agriculturalists, but herding is also a consequence of animal domestication, and animal domestication is overwhelming a result of horticultural abundance. Shepherds could be rough and lawless characters, raiding villages for booty and for fun. It's more likely that civilization, properly speaking, is an outgrowth of shepherd raids that became institutionally permanent. That is, the Cain and Abel story should be read as a tale of cultural self-justification, with the actual dynamics reversed—much as I grew up with the conviction that Indians were savage and cruel until white Euro-American civilization corralled their wickedness on bountiful reservations and made them civilized.

women into a smaller and smaller space. Feminine divinity figures declined in religious value and cultural power as warrior civilization came to control the village. Feminine divinity figures were even eliminated entirely in the new Father-only monotheisms. (At the end of *God at 2000*, the essayists/panelists convened in a group discussion. A person from the audience asked Joan Chittister, a Benedictine nun, about the "concept of goddess." Chittister responded by saying she gets "no mileage out of goddess at all. Zero." She went on to say that her "connotations of goddess" derive from her classical education, Greek in origin: "The notion of a goddess who ate her own children was not an allusion that I could live with spiritually and grow from."[7] This is a truly unfortunate answer. The "goddess" question has enormously less to do with a civilized Greek goddess who eats her children—another Cain and Abel fable?—than with an understanding of the pervasiveness of feminine divinity in the Neolithic. Chittister also says, in the question-and-answer session after her lecture, that "Ecofeminism is both a philosophy and a theology that must germinate. When I look at studies in ecology today, I have yet to see someone point to the link between patriarchy and the destruction of the planet."[8] Yet the connecting link is as plain as the nose on our faces: it is utopian male civilization "conquering" nature and impounding all noncivilized cultures as economic pawns in "development." It is also, in the Abrahamic religions, the total denial of a feminine attribute in the godhead.) Women lost religious and social power as the political, economic, and religious life of the city was increasingly controlled by men. If women gardeners, via horticulture in particular, were at the energetic and productive core of the early agrarian village, men with swords and spears lined up in the resource-controlling phalanx of extractive civilization.

The Protestant removal of the Virgin Mary (the "Mother of God") from religious veneration was a symbolic harbinger, in the early modern period, of another major loss for women, a mythic diminishment of the feminine precisely as European male civilization was unmooring from its North Atlantic bases. One of the most brutal indicators of this loss was witch burning. Less well known and more poorly understood (less understood because to understand it is to have empathy for peasant culture) was the simultaneous disruption of the traditional economic role of women in household provisioning, a disruption caused by the growing commercialization of food, clothing, shelter, and medicine as controlled and conducted by men. Virtually all consumable commodities began to pass through the economic market place before reaching the household; both the production and distribution structures were overwhelmingly in the hands of men. The public sphere was held by men; the forcing of simple necessities into the "public" market simultaneously with the undercutting of subsistence householding resulted in an enlarged masculine world and a shrunken or more constrained feminine world. The feminine role as chief provider in the traditional household economy came essentially to an end. Men, via commercial agriculture and then industrial agribusiness, provided virtually all the food for the first time in history. Women may have domesticated many of our basic foods, but the industrial and scientific revolutions generated the infusion of male control into virtually all arenas of public and private life. One stunning example stands out: in the field of medicine, male doctors were for a time the only licensed and legal practitioners of gynecology, and midwifery was abolished. Never before in human history had women been denied access to primary attendance at birth, the most fully female of all biological facts. If secret male rites of initiation in primitive society had been shaped around surrogate birth—being "born again"—then the objective scientific reductionism of modern male medicine tried to take birth itself away from the "incompetent" hands of women. (I don't think speculation here regarding "womb envy" is uncalled for.)

If the strong-arming of the agrarian village from its female founders was the first male revolution or theft or crime, at least in a civilizational sense, then the radical consolidation of economic power in the hands of men in the modern industrial and globalizing period is the completion of that assault. The industrial revolution was indeed a revolution, just as the imposition of civilization was a gendered revolution. Since Luther's time, more or less, especially in the West, male power has expanded in the public realm as never before; and the accrued crises within our present civilization are, above all, the expression of a triumphant masculine ethos.

The armed and predatory civilization of *man* has now reached global proportions; its "externalities" invoke both climate change and an apocalypse of "civilization" versus "terrorism" with Weapons of Mass Destruction looming over all. Predatory male energy made global through an incredibly technologized civilization, with fiercely contending male monotheisms providing centrifugal spin and wobble, has, with apocalyptic weaponry, created a true planetary crisis of epic dimensions. Utopian civilization is proving itself biologically unsustainable and culturally incompatible within the limitations of Creation, just as Paul Hawken says.[XV]

XV. It's a coincidence that I've been reading both *Jesus, Interrupted* by Bart Ehrman

The Kingdom of God Is Green

and *Faith and Politics* by Reinhold Niebuhr. They are unusually lucid books. Ehrman's style is simpler but very clean. Niebuhr's is more free-form but also, at times, genuinely elegant.

Ehrman, who says he was a passionate evangelical conservative in his youth, describes himself as an agnostic who came to his agnosticism not via the scholarly historical-criticism path but (page 277) because of "the problem of suffering in the world." On the facing page (276), Ehrman says his fiercest thing about God, but it's not strictly speaking about suffering in the world. It's about hell, judgment, and eternal wrath where discarded sinners are in torment forever in a place where God wants them: "What kind of never-dying eternal divine Nazi would a God like that be?" We might therefore conclude that Ehrman's agnosticism is less about suffering per se than it is about a God of absolute power who not only permits suffering on Earth but is on record of both anticipating and inflicting eternal torment onto rejected sinners.

Meanwhile, Ehrman's understanding of Jesus—a view very much arrived at via the careful process of historical-criticism—is as an apocalyptic prophet. (See especially pages 156 through 171.) Since in Ehrman's understanding a Jewish apocalyptic prophet presupposes the all-powerful God of Israel—the same God Ehrman passionately embraced in his youth but came to doubt in his adulthood—one can't help but conclude that Ehrman's agnosticism massively saps the divine energy out of Jesus. This in turn means that the "kingdom of God" is to be understood as a projection of the apocalyptic and prophetic unconditionally dependent on the will and capacity of an all-powerful God. Take away this depiction of God from Ehrman's construct and the "kingdom of God" is essentially delusional, little more than a ghost dance, as a threatened culture grasps at supernatural straws in a desperate but doomed attempt to stave off extinction.

If Bart Ehrman methodically worked his way to his present understanding via rigorous biblical scholarship, Reinhold Niebuhr seems to be something of a reincarnation of the famous Bishop of Hippo, Augustine. When Niebuhr talks of God, it's invariably of "God's order" or "God's law." God is a "jealous God" of "divine Majesty"—a "Majesty which transcends all temporal sovereignties." (Page 100.) Jesus comes off as a much tinier version of the majestic God. Jesus articulates a "law of love," but this "kingdom of perfect love . . . cannot be completely realized in history." Attempts to realize or create such a kingdom are therefore "utopian" and doomed to failure, even doomed to dreadful failure. Such attempts are sincere but misguided efforts to create "social perfectionism," as exemplified by the "radical sects" of the Reformation period. (See pages 171 and 172.)

However, Niebuhr really does have a fertile mind. So he can say, on page 27, that "The ultimate problem of myth is always the problem of God." Myth, in other words, is to a very large extent a trickle-down derivative of God-image.

So it is worth noting that there is no depth analysis of civilization in either Niebuhr or Ehrman, no embracing Spirit, no transformative "yeast," no remembrance of the precivilized Mother, no hint that Joachim provided the basis for a real historic breakthrough and broadening of overarching myth. Locked in by the biblical God, it's either weary agnosticism (Ehrman) or vigilant Augustinianism (Niebuhr). From this we can conclude that we are in desperate need of a theology that is no longer contained within civilized or utopian constraints. Only by allowing in and embracing noncivilized and eutopian dynamics will we arrive at a mythology capable of reshaping the ecocidal ramifications of "believing in" a "never-dying eternal divine Nazi" who has "mandated"—Niebuhr is rife with "mandates"—a civilizational system undulating

Male practices of predation have had a useful but limited place within hunting and gathering cultures and within the leisurely subsistence agriculture of the cooperative and communal village. A shepherd, though a possible bandit, is basically a subsistence-oriented person, too. One presumes that the raising and killing of animals for meat—primarily a male activity—will not disappear in the cooperative village to come. The human symbiosis with animals is far too old, deep, and complex to be done away with in a new sort of vegan utopianism. And while predatory hunting—the killing of animals for food, clothing, shelter, and implements—cannot be eliminated as a root of war (as animal husbandry, including nomadic herding, is a further domestication of hunting), vegetarianism does not adequately address and certainly does not resolve, by its unilateral renunciation of animal domestication, the difficulty of civilized predation. The problem is not so much the eating of meat from a flock of barnyard chickens or from a deer in the woods as it is the utopian rationalization of giant feedlots, of chemicalized and bioengineered agribusiness, of processed foods with hormones and antibiotics, of huge slaughter facilities, and of the immense distribution system represented by supermarkets and fast food. In its modern capitalist form, civilization has utopianized predation, and no technical fix or ascetic practice adequately addresses the historic depth and ecocultural magnitude of this crisis.

Religion, law, medicine, government, the military, business corporations, labor unions, scientific and educational institutions—all these and more have been, and still remain, overwhelmingly in the hands of men, although women's participation especially in religion, law, medicine, and government grows by leaps and bounds in the United states. The economic, military, and governmental policies that have brought the world to its present global crises were formulated and enacted almost entirely by men. The socialist revolutions that were to bring about the "brotherhood of man"—one thinks especially of the Soviet Union and China—were primarily the acts of men; the ensuing industrial policies of those regimes were also empowered and directed by male control, by male conceptions of what constitutes a desirable society, a "victory" over Mother Nature and all male competitors.

The women's liberation movement, arising out of an unprecedented exclusion of women from both public policy formation and productive economic life, and emerging precisely as male practices lurch us toward global catastrophe, strikes at a pattern of male hegemony at least as old

---

with diseases and traumatic institutions. It's time to go beyond civilized theology.

and as entrenched as class domination. We now have ample evidence to assert without hesitation that socialist revolution in and of itself was not, and is not, comprehensive enough to effectuate a liberated humanity. To burst through civilized male utopianism is to germinate an ecofeminist sensibility that goes to the root of Class and War and faces into the thicket of "traumatic institutions." And *that* is what makes the women's movement such a profoundly revolutionary force.

IX

Now, having established, tentatively at least, the metaphorical possibility of delineating three historic "ages" according to Mother, Father, and Son, "ages" that correlate well to social evolution, we can try to understand the potential meaning of Joachim's Age of the Holy Ghost. To gain a clearer perspective, we first must see what the Holy Ghost represented in traditional Christian theology.

We have already seen that the Holy Ghost was defined as the Spirit created by the reciprocal love between Father and Son. In the *Encyclopaedia Britannica*, in a long article on Christianity, the Holy Ghost (or Holy Spirit) is characterized as a "dove," the "life-creating power," the force of "creation and rebirth," and spontaneous, charismatic overflowing. One can easily see that these are attributes traditionally associated, not with the masculine, but with the feminine.

Mary Daly, once a leading radical theologian, used Joachim's schema of three ages as a basis for an attack on Christian theophany. In her *Gyn/Ecology*, Daly insisted that the trinitarian godhead is entirely the work of male consciousness and that all three "persons" in the Trinity are masculine figures. In that assertion, Daly was being entirely consistent with traditional Christian theology.[XVI] But Daly also referred to the Holy Ghost

---

XVI. Carter Heywood, an Episcopalian and one of the first women priests in the United States, says on page 229 in her book *Our Passion for Justice: Images of Power, Sexuality, and Liberation* that "one of the most difficult tensions Christian feminists encounter is between our life experience as feminists and the doctrine, discipline, and worship of a church founded on masculinist assumptions about the relation of a Father God to his Son Jesus who himself is God the Son, and who together reign in a Kingdom, in which power is 'naturally' handed down from above—and only to those who submit to the omnipotence and omniscience of a Father who knows best. The Creator and Redeemer are experienced and conceptualized, explicitly, as being in the image of *men*. This is no historical accident, no metaphysical coincidence. Indeed, in the image of the prototypical and primary male-male relation (father-son), these doctrines are constructed and sustained/sanctified (by the third male-Spirit, the one who sustains

as a "mythic male mother" and, in a chapter called "American Gynecology: Gynocide by the Holy Ghosts of Medicine and Therapy," she calls Joachim's third epoch the "Age of Gynecology."

Yet Daly's own data contradict her interpretation. In discussing the etymology of the name Lilith—Adam's original wife, according to ancient Hebrew myth—Daly notes that "Lilith" derives from the "Babylonian-Assyrian word *lilitu*, meaning a 'female demon, or wind-spirit.'"[9] Daly then goes on to show how that definition is similar to the definition of the Holy Spirit—the latter term grounded in the Greek *pneuma*, meaning wind or air.

Since we are in the realm of cultural myth—even in the ethereal metaphysics of theology!—there is no "proving" anything with empiric finality. Metaphor is beyond positivist proof. (That fact—metaphor is beyond positivist proof—is also one of the reasons a large number of people cling to empirical science. If one is committed to empiric reductionism, metaphor can't help but be a slippery slope leading one down to the bog of unsubstantiated imagination. The dilemma is compounded and made more deeply ambiguous in that a great deal of empirical science is financially channeled according to the dictates of aggression and domination. This makes the "empiricism" of science subservient to pre-established political and economic goals, creating a kind of "scientific" tunnel vision that causes science itself to become, as enabler to civilized utopianism, an auxiliary myth. Science, with its clean numbers and antiseptic formulas, with its bright knowledge that mimics and taunts wisdom, can blithely wear a patriotic white hat, a Boy Scout badge. Science, in its multiple techniques, has pried extensively into the inner workings of Creation, into the secrets of nature; but the core energy of science, at least in its utopian impoundment, is a pseudoinnocent boyish immaturity, never doubting its right to experimentation, outraged by any thwarting, and contemptuous of "girlish" emotionality.)

The best we can do, without conclusive proof, is attempt to grasp the buried psychological, cultural, and spiritual meanings in myth and project them in a way that lends intellectual clarity and mythic coherence to our actual lives and present conditions. Myth must provide us with a worldview that "explains" the meaning of our lives in the world and our trajectory in history. To be sure, we must proceed with a great deal of spiritual caution and intellectual restraint, inviting challenge and rebuke;

---

and sanctifies) to benefit primarily *men* in their efforts to live meaningfully and well on earth."

but simply to dismiss mythic conceptions as utterly meaningless is itself a kind of blind cultural chauvinism. To assert that mythic symbolism has only been deliberately and consciously employed as a devious tool by sex or class elites to hold women in gender inferiority or the lower classes in economic subjugation (especially the peasantry, in its long history prior to the industrial revolution) is also a kind of reductionism, although one with which I can sympathize. These images of male superiority were and are culturally entrenched. Their entrenchment lies deeper than mere existential calculation of advantage—that is, they became utterly normative—though it can legitimately be asserted that the historical longevity of these images derives incalculably and exactly from their male and upper-class advantages. One need not believe in their theological literalness to understand and appreciate the symbolic dynamism or to realize that male mythology will crumble with astonishing speed—not unlike the collapse of the Soviet Union—once male hegemony is broken, even if we can now hardly imagine such a world or what it might take to reach such a world.

If we can even tentatively accept the casting of "ages" in human history since the dawn of horticulture—and remember that the "matriarchy" of "primitive communism" propounded by Engels and Marx was based on Bachofen's *Mother Right*—then we have to proceed in a way that is *psychologically* consistent. The unfolding of the cultural unconscious has its own logic. If we can identify an "age" of the Mother, an "age" of the Father, an "age" of the Son, then the next "age" simply has to be the Age of the Daughter. Psychological consistency demands it.

This mythic formulation thus "accounts" for (enables us to explain to ourselves) the remarkable emergence of a global women's movement in the extreme cultural fracture of industrial civilization, in the planetary toxicity of "man's mastery over nature." It explains the cultural and psychological need that prompts some radical feminists to try and revive old goddess religions—although, again, one needs to go beyond Greek goddesses who, Chittister says, supposedly eat their children, to the broader and deeper embeddedness of feminine divinity in the ancient agrarian village, and perhaps even earlier. (Unless, of course, one truly believes that feminine divinity is inherently cannibalistic and that masculine divinity has mercifully put a stop to female blood lust, comparable to how civilized Euro-Americans rescued Africans from paganism and Indians from savagery.) This is not a flimflam exercise of invoking "fertility" but of recognizing what Marcus Borg calls the "womblike" compassion of divine character.[10] It exonerates the growing disaffiliation of people from conventional religious forms and expressions that refuse to change or look beyond myopic horizons, most certainly the myopic horizon associated with the rise of civilization. And because women have traditionally, culturally, and "unconsciously" been associated with the Earth—*Mother* Earth, *Mother* Nature—the formulation helps explore the growing power of all forms of ecological sensitivity as a feminine impulse and helps locate the sexual symbolism within that emerging ecological sensitivity.

By unconscious symbolic association, the liberation of women requires the "liberation" of nature and the Earth. Women's liberation implies the end of civilization as we know it and the resurrection of rural life on a new and finer plane. And if it can be argued that women's liberation does not imply the end of civilization, then it most certainly does imply the turning of civilized aggression inside out, a process whereby the spiritually cleansed institutions of civilization—i.e., no longer traumatic, no longer rooted in Class and War—could serve the ends of ecological sharing rather than unlimited and rapacious taking.

Let me hasten to affirm, once again, that we are no longer in the realm of pure empirical study or of precise empirical verification. This is, rather, the poetic turf of psychoanalytical association, of mythic imagery, the stuff of the unconscious brought to consciousness in a period of intense political, cultural, religious, and natural stress. But the evidence is all around us and is, in fact, beating on our doors. We might even call it prophetic.

The women's movement can be more fully understood in its cultural complexities by casting it against the immediate background of masculine industrialism. Simultaneously with the industrial revolution there was growth in human population generally and also a mass migration, compelled by economic strictures, from the countryside to the cities and from the Old World to the New World. *Hungry for Profit: The Agribusiness Threat to Farmers, Food, and the Environment* is a book of thirteen essays published in 2000 by Monthly Review Press. In the first essay ("The Agrarian Origins of Capitalism"), Ellen Meiksins Wood insists that modern capitalism was created in the English countryside, by the landed English ruling class, before the start of the industrial revolution. Though narrowly this analysis seems true—"capitalism" as a market system is a relatively modern conception—it neglects to emphasize that, as a wealth extraction system designed to concentrate wealth in few and fewer hands, capitalism mimics and extends the aristocratic or upper-class control and concentration of wealth, a process that simply saturates the history of all civilizations. Capitalism, once pried loose from explicit aristocratic control but still operating with civilized values, results in the "democratization" of economic exploitation. That capitalism devised or discovered radically new techniques to extract wealth—by penetrating and destroying rural culture as never before, by penetrating and manipulating nature as never before—does not mean in the least that its core civilized motive has been altered whatsoever. As Lewis Mumford shows in "Utopia, the City, and the Machine," the utopian impulse to surpass both nature and all cultures embedded in nature is coeval with the rise of male civilization. Civilization simply had to wait until industrial capitalism provided the techniques by which nature could be fully "conquered" and the peasantry evicted permanently from the commons.

In the eighth essay of *Hungry for Profit*, Farshad Araghi supplies statistics for what he calls the "emptying of the world's countryside": in 1800, he says, 98 percent of the world's population was rural, 70 percent in 1950, and in the year 2000 only 55 percent.[11] The spread of capitalism worldwide has resulted (England, with its industrial revolution, led the way) with landlessness for peasants, poverty and homelessness, the destruction of subsistence, and the growth of what Araghi calls "megacities," with ten million or more inhabitants. "At the beginning of the [twentieth] century," writes Araghi, "the vast majority of the world's peoples were peasants or lived in rural areas. By the end of the century, peasant life was rapidly disappearing and the world had become overwhelmingly urban."[12] Utopian predation drives indigenous people, peasants, and all other subsistence-oriented persons out of Creation and into the hypertechnologized City of Man, from reciprocal, biological interdependence into privatized, electronic destitution.

Dislodged from the household economy, an economy largely agricultural and in many ways cooperative if not overtly communal, women and men found their cultural environment disintegrating around them. Thrown out of the garden and into the maw of the atomizing, Hobbesian, urban-industrial, capitalist job market—selling labor for survival, literally wage slaves—both women and men began to slide into social isolation and individualized anomie. The extended family became unglued. The nuclear family emerged. "Culture" was what one watched on TV or bought at the mall.[XVII]

XVII. There is no really good term here, so I'll say "householder." All hunter and gathering peoples were householders, as were all peasants and most small-scale farmers. Householding means a pattern of subsistence and self-provisioning behaviors that constitute a life rhythm in a very familiar rural environment, a life with food crops, animal husbandry, hut construction, firewood gathering, and so on. The character

Age of the Daughter

In the utopianized context, lacking traditional community support and the self-sufficiency of the household economy, women slowly but steadily began to analyze and rebel against the isolation, vulnerabilities, and chronic injustices in their lives, both at home and in the larger economy. (Men groped desperately for fraternity in unions, from which women were pretty thoroughly excluded—the men reverting, as usual, to the boys' club.) As the urban-industrial atomizing process advanced, so too did the women's movement grow, albeit in fits and starts. Now, in the last five or six decades, as the masculine military-industrial machine has forced its utopian agenda into every nook and cranny in the world, and as certain very angry men strike back in the name of Allah, a global women's movement is assuming unanticipated proportions, reaching right into the United Nations. If male civilization in its industrial mode has undone eutopian householding, a certain (as yet insufficiently influential) women's power is reaching to transform utopian policy and politics.

These phenomena underscore the growing crisis within the masculine ethos and the increasing readiness of a new feminine ethos to participate fully in a new and finer level of political and economic determination. The Age of the Daughter is rising to confront—and to transform—the threat of global political, ecological, and spiritual catastrophe. The dove of peace is awakening. The "consecrated anarchy" of ecological coherence is socially alive. The women's movement is not only the promise of the future; the women's movement is the measure of whether there will even be a future.

X

Let me add one last, and I think rather striking, example of this symbolism. Here in Wisconsin, in the Capitol in Madison, the high-domed rotunda rests in classical style on four "galleries" that house, respectively, the Supreme Court, the Senate, the Assembly, and the Public Hearing Room. Between the galleries, atop the high pillared walls, are embedded four massive mosaics representing Justice, Legislation, Government, and Liberty. These—"abstractions," perhaps—are depicted in human form, an

---

structures associated with this ancient heritage and inheritance have atrophied in the West to the point of near oblivion. Utopia, as it were, has extinguished them. Yet a eutopian world requires the resurrection of householding, with many of its folk attributes. This also requires the rediscovery of our physical bodies as we learn anew craft competence.

anthropomorphism not only common but probably necessary for human understanding.

Justice is represented as a dispassionate matron, holding the scales of justice in her hands. *Legislation* is a Blakean patriarch, beard flowing, holding in one hand a tablet and in the other a *stilus*, an instrument for writing. *Government*, with a sheathed sword in his left hand and a baton (or tightly rolled, secret fraternal message) in his right, is dressed as a warrior, after the Greek or Roman fashion. *Liberty*, with a garland in her hair, is a voluptuous maiden, the shoulders of her gown falling down upon her arms; with her right hand she protects or upholds a globe—the world—the Earth—and with her left she points upward toward the sky. Here, in full public view, are the Four Ages in magnificent depiction, available to the understanding of all those with eyes to see. And this proves, it seems to me, that this "myth" has been patiently waiting for the awakening of human discernment. So we are talking about *recognition*, not *fabrication*.

Can't we now say, with all the evidence that both historical scholarship and modern depth psychology provide, that the "primitive communism" of the early agrarian village (however faulty such a designation may be) did give birth to a new form of social justice? That the rise of the patriarchal city did result in the first sustained effort at formal legislation? That the Greek and Roman experiences in (very limited) fraternal democracy did provide new forms of government waiting to be extended to all? And can't we also say that the women's movement does promise the ample growth of both personal liberty and ecological integrity—a eutopian "restoration"?

The trends, however qualified, are there for all to consider. What's needed is an overarching myth. Wisconsin already has it magnificently embedded in the rotunda of its Capitol.

XI

Having raised the question of myth, perhaps we can now indulge in some speculative thoughts on what will help bring about a "new age." The failure of the New Deal and Great Society programs to provide comprehensive economic democracy and sustained social equality has pretty well discredited the American liberal establishment—or, at least, the democratic energy unleashed by the New Deal especially was not sufficiently deepened and revitalized in subsequent post-war decades. What's left of that energy is now under severe attack and strain. Kennedy, Johnson, and

Humphrey were all entangled in the Vietnam War. McGovern's apologetic mildness, coming after institutional gains in civil rights under Johnson, further disgusted the white male working class. Carter's Sunday school moralisms caused voters to choose a Hollywood king just as England had chosen a nostalgic Empire queen—King Ronald and Queen Margaret. (Both countries have since had an amazing series of testosterone princes.)

Unable or unwilling to separate themselves from the underlying male, capitalist worldview, liberals have been incapable of formulating an adequate alternative or a more comprehensive policy. The dominant worldview consists, first, of a virtually inescapable presumption of "civilized values" and, second, a psychology of male superiority generated from both civilized and religious sources. All this is compounded by corporate control of the media and the psychological emasculation of all who criticize or attempt to bring under control the male "warriors" with their weapons of apocalypse. The best of the liberals, in the political sphere, have become the intimidated yet bravely undeterred little boys with thick glasses and artistic temperaments in a bleak schoolyard filled with jocks and bullies. The kingdom of God continues to be jeered at and abused for its "sissy" advocacy of servanthood and stewardship, for its willingness to attentively listen and quietly respond.

"Science must teach us the future," said Walter Mondale, Democratic Party candidate for president in 1984 and rejected surrogate for Senator Paul Wellstone in 2002. Deference to technocratic development and strategic advantage has become the leading political principle, and that entails an avoidance of engaging real issues and an evasion of both sharing and stewardship. The conservative Right has repeatedly swept into power with sweeping proposals to brush aside a great body of liberal legislation and to allow (theoretically) the industrial private sector to provide the ways and means toward abundance and affluence through increased productivity at the expense of economic and environmental regulation. The 1990s saw an astonishing magnification of wealth concentration, wage stagnation, agricultural depression, and fantastic mergers. The bulk of this happened in the administration of "the first black president," Bill Clinton, whose charming public persona projected virtually the total opposite of his actual economic policies, from NAFTA to the "war on drugs" to the dismantling of welfare. With Bush junior, initially "elected" by the Supreme Court, we entered, after September 11, 2001, the death dance of Western civilization. And with Barack Obama, we see the lofty rhetoric and endless temporizing of a "liberalism" nearly swallowed by the voracious monster of capitalist empire, both in terms of corporate bondage and of international militarism.

For the time being, the American people appear to be emotionally milling around, ambivalently weighing the punitive policies and the continued shrinkage of social programs compounded with a diffuse fear of terrorism and an ambiguous yearning for homeland security. Witness the sharply divided electorate in 2004 and Bush's narrow electoral victory or the (misguided) enthusiasm for Obama in 2008 followed two years later by an upsurge of (misguided) Tea Party outrage. Liberals are apologetic and pleading, offering neither a comprehensive alternative nor any sustained resistance to the reactionary onslaught. Liberals can't bring themselves to believe in "conservative" verities even as they are lacking in any substantial alternative vision, at least any they are willing to say clearly out loud. Needless to say, the public has not yet felt the full consequences of the policies it presently tolerates—"drill, baby, drill"—just as the world as a whole has yet to feel the full impact of the war on terrorism or climate change. As those realities go deeper and deeper, as recognition of them (climate change especially) is no longer avoidable, we are likely to witness an unprecedented radicalization of the world's women—although the alternative in the indefinite short-term is right-wing "warrior" control, with all the chest-thumping denial that goes with it. And the Right, we always need to remember, has an ideology rooted in control. Control is what the Right does best or at least reflexively. It has little hesitation to impose order, ruthlessly if deemed necessary.

The women's movement represents an awakening of women to their dormant political strength, to a set of ethical values and emotional strengths anciently rooted in women's experience. Even though it is not yet ideologically (or mythically) coherent, the women's movement in its deepest conviction is in radical opposition to the utopian system of male control. Chittister again: "A feminist value system is other than a patriarchal value system."[13] As the male civilized war against male terrorism becomes bloodier and bloodier, the disgust of women will congeal in political outrage; and then the male system will be overthrown.

But women are not the only "group" whose radicalization is at hand. Minorities, old people on fixed incomes, the handicapped and disabled, college students, blue- and pink-collar workers, farmers and loggers—all must eventually realize that the wealth and prosperity promised by "Reaganomics," "globalization," and "free trade" falls heavily into the hands of the already wealthy, and that Creation—Earth, Air, Water, and Life—cannot be continually abused, taken for granted, and ignored. Republican policymakers of the last thirty or forty years have been cheerfully planting the political depth charges that will eventually blow them completely out of the monopolistic electoral water. Yet is it not appropriate for the dysfunctional capitalist globalized military machine to be governed in its decline and perhaps even collapse by that party holding most closely to traditional capitalist ideology? History has its own jokes to play, its own ironic poetry.

This is, however, no time for smugness. An ideology of total control uncontrollably losing control is not a pretty thing. The decline of capitalism—and of utopian civilization generally—is an extraordinarily dangerous phenomenon. Rural culture crumbles everywhere, as Farshad Araghi's statistics show. And that means that the real "safety net" of embedded rural culture, with gardens, chicken coops, goats and forest, has been shredded. A billion people, says Araghi, do not have adequate shelter. A billion rural people are landless or near-landless. People are dying violent deaths all over the world. Cities teem with misery.

Over the last six decades the United States has maintained a permanent war economy and has steadily become the major arms supplier in the world. But even here there is a tragic irony. It was World War II that pulled the industrial world out of the Great Depression, with America leading the way to "prosperity" and its primary beneficiary; yet reliance on military prowess and Manifest Destiny will prove to be among the millstones that pull capitalism down. The major political dangers that lie ahead are global war and the growth of an explicit fascism—not to speak of our sheer political stupidity in regard to climate change and peak oil.[XVIII] Never before has the need been so great to put before the American people—the world's people—an alternative political and economic vision that unhesitatingly utilizes spiritual ethics (servanthood and stewardship) as the foundation for a new, more equitable world. The questions are: who will do it, and of what will it consist?

Within the next decades we will very likely witness the emergence of a genuinely radical international political movement with broad public support. Whether that movement will be embraced by the Democratic

---

XVIII. In the May 21, 2012, issue of *The Nation*, Alexander Cockburn in an article on page 9 called "So Who's the Fascist Here?" says that "Fascism, among other things, is a system of extreme, methodical state repression, violent in contour and threat, buttressed by ultra-nationalist mythology, a militarist culture and imperial ambition." After pointing out that fascist regimes repress labor, are the sworn foe of the right to assembly, and spy obsessively on citizens, Cockburn insists that "We live in a fascist country—'proto-fascist' if you want to allay public disquiet, though there's scant sign that most Americans are disturbed by the trends."

Party remains to be seen. (Ralph Nader, as presidential candidate, was unable to drive either Al Gore or John Kerry to the Green Left, which was the direction needed to decisively beat Bush—or, if not to beat Bush in those specific elections, then to articulate a body of coherent policy sufficiently full of truth and substance and therefore capable of shaping a new and far more comprehensive policy base for future elections. But short-term calculation, no matter how myopic, always seems to "win.") Because the Democratic Party is also closely tied to the capitalist, industrial, and financial worldview renders its Greening ambiguous; and now the fear of being associated in any way with "terrorist" grievance—even to acknowledge that there may be real grievance—has made political timidity into a holy grail. But wherever it comes from, a powerful radicalization will certainly take shape, and through it women will finally rise to the highest political offices and begin to shape a new and radically different public policy.

Yet the real social transformation is far more complex than having a woman sit in the oval office. The real revolution involves the end of terrorism (using Noam Chomsky's definition), the demilitarization of American foreign policy, and the disarming of every economy. It involves a range of experiments in decentralized, ecological socialism even as it empowers a democratically restructured United Nations. It involves big reductions in energy production and consumption and the reconstruction of passenger rail service. It involves the phasing out of chemicalized agribusiness and the repopulating of the countryside with cooperative land trust organic gardening and farming. It involves the resurrection of householding and a renewal of folk culture. It involves much more naturalness and far less compulsion in our schools. It involves greater participation in international cultural exchange and the steady encouragement of interfaith dialogue. (Meditative Buddhism—or its disciplines—may well blossom in the West.) In the process, we must hold fast to and be guided by the principles of sexual reconciliation and racial equality, ecological coherence, and socialist sharing.

Through all this there will be the growth of a new cultural and religious myth. This "new" myth will enable Christians to discover the peasant Jesus hidden in and buried under traditional utopian orthodoxy. (They will find the true Son locked in the dungeon of the "Constantinian Arrangement." Those who love Gospel teaching will also find the true Mother in the dungeon of civilization itself, locked out of the orthodox Trinity, perhaps assigned to the cellblock for pagans.) A passion for peace will come with disgust over war.

Barbara Ehrenreich, in the final remarks of her *Blood Rites: Origins and History of the Passions of War*, says the "passions we bring to war can be brought just as well to the struggle *against* war":

> What we are called to is, in fact, a kind of war. We will need "armies," or at least networks of committed activists willing to act in concert when necessary, to oppose force with numbers, and passion with forebearance and reason. We will need leaders—not a handful of generals but huge numbers of individuals able to take the initiative to educate, inspire, and rally others. We will need strategies and cunning, ways of assessing the "enemy's" strength and sketching out the way ahead. And even with all that, the struggle will be enormously costly. Those who fight war on this war-ridden planet must prepare themselves to lose battle after battle and still fight on, to lose security, comfort, position, even life.
>
> But what have all the millennia of warfare prepared us for, if not this Armageddon fought, once more, against a predator beast?[14]

This "predator beast" is *within* utopian male civilization, and is in the end protected by all religions that require totalitarian Father worship. This current war, a self-described war of civilization versus terrorism, essentially announces a fight to the finish between the three Abrahamic brothers—an end to war, an end to brotherhood—even as the so-called "Arab Spring," the huge public protests in Europe over E.U. "austerity" measures, the Wisconsin "uprising," and the Occupy movement may well be indicative of "committed activists willing to act in concert," prepared to lose but struggle on toward a hope that burns within.

If myth enables us to focus our energy in the direction of eutopia and promote the cause of peace, then we must learn the value of myth, or at least cease being so immediately hostile to it. If myth enables the establishing of dialogue, then myth will have rendered the humane goals of dialogue and integration an invaluable service. But to get to a place of mythic peace, we first have to confront the "sacred violence" that undergirds and protects our dominant myths—the myths of civilizational supremacy and the strict masculine nature of the divine.

No human culture has ever existed without myth. No human culture ever can exist without myth. If we are too proud to wrestle with the intellectual dilemmas of myth, we can be sure the Right can supply and will promote its own, however canned, brittle, irrational, and rigid. The effort to live without myth is doomed to failure: even "scientific" Marxism had to rely on myth—from the alpha of "primitive communism" to the omega of the "withered state." Myth provides the social groundedness and cultural visions without which we are disoriented and lost, unable to project cooperative or communal purpose. The question of myth is not confined to religious belief or even to incidences of "founding violence." Myth is a cultural construct with powerful political implications.

This emerging "new age," the Age of the Daughter, will not come into being without the formation of an explicit and overarching myth that embraces the feminine and exposes "sacred violence" for what it is and what it has done.. But first we have to demythologize our empirical illusions of endless technological progress and ceaseless affluence. We need to dissolve the rigidities within fundamentalism, especially the (in the end disastrous) image of God as psychotic control freak, a manic "Father" loaded with cosmic weapons who may burst in upon us at any moment and blast us all (unless we can quickly produce our belief ticket) into a torment of fiery, eternal hell: Bart Ehrman's "never-dying eternal divine Nazi."

The future is clear—if we can get there. The "kingdom of God" is, always, fully and compassionately at hand. The question is whether we have, collectively, the humility, courage, and reverence to abandon our religious salvation illusions and utopian economic addictions, to let go our transcendent fear of the Father God and discover, instead, the endless immanent love of the Mother Goddess.

The crisis we are already in, and into which we will all become more deeply thrust, involves the crushing of gender stereotypes, including those gender stereotypes blown into religious images by the psychopathology of civilized history. Clinging to conventional identity as male and female—especially and particularly male—is to be drawn irresistibly into this crushing, into this conflict, into this carnage. Those who voluntarily submit their identities to spiritual transformation (and it is possible to do this from within every major religious tradition) can resist or evade this crushing and begin to both envision and experience what human life looks like on the far side of utopian collapse.

A "mythic" understanding of real history—successive "ages" of Mother, Father, Son, and Daughter—provides us with a comprehensive story of human life on Earth since the last Ice Age. It gives us a wholesome future to believe in and work toward. Spiritual transformation depends on our willingness to submit to humility and reverence, to sharing and stewardship, and to a far more balanced composition of our sexual identity.

Spirit, as always, is in and through and beyond all this—Mother, Father, Son, Daughter, and who knows what else. Our task is to be faithful, loving creatures on this magnificent planet.

Why is that so hard?

Notes

1. Merton, *Faith*, 219.
2. Block, "The Right's," 20–21.
3. Roy, "Fascism's," 18.
4. Nichols, "Dark," 32.
5. Mitchell, *Gospel*, 5.
6. Lewis, *That Hideous*, 316.
7. Chittister, *God*, 148.
8. Chittister, *God*, 70.
9. Daly, *Gyn/Ecology*, 86.
10. Borg, *God*, 11.
11. Araghi, *Hungry*, 145.
12. Araghi, *Hungry*, 158.
13. Chittister, *God*, 70–71.
14. Ehrenreich, *Blood*, 240–41.

24

Stuff

MARCUS BORG HAS TOLD us that, in preindustrial agrarian society, the top one percent of the urban ruling elite (the ruler and governing class) consumed "about half" of the total economic income or production. The top ten percent (including not only the ruler and governing class but also the "retainers," well-to-do merchants, and the upper echelon of the priesthood) got about two-thirds of the income or production. That leaves—to do the math for those who may have forgotten basic addition and subtraction—one-third of the income or production for the ninety percent who produced, roughly, one-hundred percent of that income. And *that* is the template for the relationship between the "dominant minority" and the peasants.

Whether one reads Ralph Nader in *Crashing the Party* ("When three hundred of the richest people on Earth have wealth equal to the bottom three billion people on Earth, extreme affluence is built on the backs of extreme mass poverty"),[1] or Frances Moore Lappé and her daughter Anna Lappé in their book *Hope's Edge* ("The pay of America's CEOs leapt 535 percent during the '90s, while most working people barely kept ahead of inflation. And those who now control as much wealth as half the world's people could fit into Anna's high school auditorium")[2], or Kevin Phillips in his *Wealth and Democracy* ("The logic became clearer in 1999, when the *New York Times* began a news story with a stunning new truth: 'The gap between rich and poor has grown into an economic chasm so wide that this year the richest 2.7 million Americans, the top one percent, will have as many after-tax dollars to spend as the bottom 100 million'")[3], one cannot avoid coming to the startling and brutal realization that, despite all the rhetoric and hype about equality and democracy, the basic pattern of economic control, of how civilization really works, of how wealth gets consolidated, has not changed one iota since the time of Jesus.[I] With one exception—no, two exceptions—no, three exceptions.

Exception one: there is a hugely greater volume of *stuff* in the economic production system, some of it incredibly toxic.

Exception two: there is a hugely greater human population, its subsistence culture largely broken, to suffer the agonies of exploitation, neglect, and deprivation.

Exception three: the enforced globalization of civilized inequality, given the "externalities" of commodity production and waste disposal, the apocalyptic nature of civilized weaponry and the magnitude of human population, has brought civilization to an intersection—a very speedy and dangerous intersection—with its fundamental falsity, contradictions, and criminality. September 11 was one such intersection.

So far as I can see, there are only three major options ahead: apocalyptic breakdown involving military showdowns, economic collapse, and ecological catastrophe; a locked-in "steady state" global aristocracy with a focused, mean, and exceedingly ruthless "rapid deployment force" that keeps all localized protest under careful surveillance and all rebellion preventively squashed; or the breaking in of the kingdom of God with its fully committed radical sharing and its reverent revolutionary stewardship.[II]

I. See also the wage and income graphs—pages 27, 33, and 38—in David Cay Johnston's 2003 book *Perfectly Legal*.

II. The bulk of these essays, as I say in the Introduction, were written in the years immediately after 9/11, and this manuscript as a whole has had to patiently wait its turn to be printed. In the meantime, I've often felt dismay over the apparent lack of recognition on the Left that civilization's globalization simply has to result primarily in one of these three options: apocalyptic breakdown, aristocratic restoration, or what I'm calling here the breaking in of the kingdom of God.

But, finally, in *The Nation* for November 28, 2011, there was an excellent article by Naomi Klein called "Capitalism vs. the Climate"—I am using a portion of it for one of this book's epigraphs—that clearly advocates for "an alternative worldview to rival the one at the heart of the ecological crisis—this time, embedded in interdependence rather than hyper-individualism, reciprocity rather than dominance and cooperation rather than hierarchy." This is exactly right.

My criticism of Klein's diagnosis, however, is that she fails to carry her etiology of capitalism—her tracing of causes—anywhere back far enough. This failure may be typical in journalistic writing, but it's also a major flaw that results in analytical myopia. The Left is simply going to have to face recognizing civilization for what it is. Part of cure is adequate diagnosis, and adequate diagnosis (at least in this case) requires as complete and thorough an etiology as is humanly possible. To let civilization conceptually off the hook is essentially to say that there are things or concepts or social

The Kingdom of God Is Green

Now, perhaps as never before, we need to really think about what we're asking when we pray "Your kingdom come." Shall we have a brutal and self-indulgent aristocracy of empire consumption or a gentle and compassionate village of God? Whatever we choose, it's helpful to remember that we are voting with our lives.

---

constructs too big to fail or too sacrosanct to ethically expose.

But let me also say that I'm not by any means stuck on calling this "alternative worldview" by the name "kingdom of God." I suppose I feel myself writing primarily for bewildered Christians (or people raised Christian who are now simply bewildered), even as I remain convinced that transformation without the mysterious, hidden workings of underlying Spirit is simply not plausible. I do not believe in the ultimate wisdom of pure human will, whether Right or Left. If our predicament were to be corrected purely on the basis of human intelligence, we'd have already achieved that end. It's not, exactly, that we're lacking in brain power. What we're lacking in and resistant to is humility and reverence. But these are spiritual qualities; they are not, strictly speaking, derivative from brain power. Intelligence is necessary but not primary.

No proof is required to say that human intelligence is an evolutionary and collective phenomenon. It's so self-evident it can go unremarked. Spirituality is similar but not identical. We humans store and transmit the great bulk of our collective intelligence in a variety of ways, even as the storing and transmitting of our spirituality is much more elusive and harder to pin down. And while both intelligence and spirituality reside, we might say, in human consciousness, it can also be said that our resistance to the spiritual, to the unfolding of its special wisdom to a far greater extent in our collective lives, is linked to our human pride in the supposed superiority of intelligence. We don't wish to let go of our sense of control. We resist feeling (or being) vulnerable.

Saying all this does not "prove" that Spirit "exists." Proof belongs to the realm of intelligence, and its investigations are not to be scorned. Science is made of such investigative stuff. The finer wisdom of spiritual vulnerability cannot be empirically proven—except by the lovely eutopian world collective vulnerability would, in my estimation, inevitably reveal. But such an "experiment" in servanthood and stewardship hinges only secondarily on intelligence or "common sense." Its true pivot point is rooted in spiritual openness. In a way, our intelligence constitutes our richness, so it is terribly hard for a rich (i.e., intelligent) person or group or country to go through the eye of the spiritual needle. It may not be just about losing control, but it is heavily about losing control—and intelligence is very much about the building and maintenance of control, even if the likely end result—as we can now contemplate—is breakdown, immense suffering, and the restoration of a primal chaos.

This means we have reached a point in the human journey where we are, via global crises, brought face to face not only with the limitations of intelligence but with the far more disturbing realization that intelligence devoid of spiritual humility is inevitably and destructively hubristic. And this most certainly includes the policy projections of political candidates (I write this in mid-December 2011, two weeks before the first Republican presidential primary—or caucus—in Iowa) who now utter the most preposterous, fantastical things in behalf of an ideology, a conviction, that asserts the limitless supremacy of American intelligence in a state of reflexive denial. It is, in its own way, an awesome demonstration of how removed we are from reverence for and in Spirit, even as all candidates compete furiously in the hobbyhorse race for the brass hat of Commander in Chief.

Notes

1. Nader, *Crashing*, 314.
2. Lappé and Lappé, *Hope's*, 7.
3. Phillips, *Wealth*, 103.

25

A Broken Bone

BECAUSE MY WIFE SUSANNA and I are folk musicians (as a young woman in Switzerland, Susanna made her own violin at a school in Brienz), we are sometimes asked for "special music" at area church services. As a rule, we respond positively to these requests for our—I should speak only for myself—for my frustration with conventional Christianity does not, I hope, mean I have developed a knee-jerk hostility to anything and everything churchy. Yet I am amazed how often I hear things, especially in sermons, that not only reinforce my criticism of the church but even underscore it. I had one of those moments this morning.

The subject—the Gospel reading—was about the rewards for faithfulness and kindness, Matthew 10:40–42, even for giving a cup of cold water to someone in need. It was hot, there were fans, and I missed (with all the white noise) some of what the preacher said. But I caught this—The great anthropologist Margaret Mead, he said, claimed to have discovered "the first sign of civilization" in the excavation of an ancient human bone that showed it had been set, and healed, after having been broken.

Later, when I asked the minister where he had obtained that story, he said "Off the Internet." So, whether Margaret Mead actually said such a conventionally silly and ethically preposterous thing remains blissfully in doubt.

When, in an extremely condensed and shorthand way, I expressed my disbelief in the linkage of a healed bone with civilization, the minister said (I had never doubted this) all he really was trying to convey was the importance of caring, healing, and compassion.

A Broken Bone

Coffee hour is a poor occasion for a seminar on history and the meaning of words. But I told the pastor that civilization, in my understanding, is the locus of our actual religion and that Christianity is, largely, a one-morning-a-week insurance policy with the possibility of an afterlife dividend in mind. The minister (he was smiling) called me a "cynic."

In conventional usage, a cynic is a person who says corrosively bitter things, emerging out of a stance, perhaps a disposition, of resentment, frustration, and anger, things overstated and therefore, on balance, untrue. Well, perhaps the shoe fits. I hope not—but that, too, isn't the main point. The main point is that, from the pulpit, a gentle and well-meaning mainstream Protestant pastor could tell a little Margaret Mead story about a healed femur, link that healed femur with the origins of civilization, and then be a little miffed when challenged (I think gently—perhaps not) on his use of concepts.

Well, what *is* the origin or the basis of compassion? Does civilization even get to posture politically with a healed femur? In the world of animals, I have seen innumerable instances of compassion, a great many of these instances—perhaps the bulk—associated with adult animals, particularly the moms, caring for their young. In yet one more of the many poignant images employed by Jesus, he bemoaned the refusal of "the chicks" to gather under his hennish wings (Matthew 23:37). The simile spoke to every peasant listener.

Compassion to some degree is abundantly built into us, but selfishness acts as if scarcity is the norm. Now it's no doubt true that selfishness predates the rise of civilization—civilization didn't invent it—but civilization with its "diseases" of Class and War *institutionalized* selfishness, putting agricultural abundance in an extractive, aristocratic vice. Civilization is the armed few (armed with weapons, God, and law) creating and maintaining a reign of imposed scarcity for the many—or, in our time and circumstance, using globalized industrial surfeit, through the market system, to make themselves unimaginably rich.

Civilization killed Jesus. One would think the church would know this, that pastors-in-training are helped to delve deeply into the conflicting values of church and state, spiritual humility versus civilizational hubris, and that clarifying these values for the laity would be high on the pulpit's priorities—always the *seeking* for the realization of the kingdom of God and an expounding of those practices, ideas, and policies that are in accord with that realization.

The Kingdom of God Is Green

But that a reasonably well-educated, kind, and gentle pastor can say, from the pulpit, that a healed femur is the origin of civilization is casual proof of how deeply the church is infected with civilized mythology. What *are* these pastors taught in seminary?

Let me add one other thought, a related one. A good friend, another pastor but in a different denomination, was essentially booted out of a local congregation. "Booted out" is somewhat ambiguous for the ruling authorities left few tracks and my friend was led to believe that resistance on his part (not to speak of a real honest-to-goodness, let-it-all-hang-out fight) would be extremely prejudicial to his career. He has a fine wife, two kids still in school, and he chose to go peacefully—though it's somewhat sordid to abuse the word "peace" in this way.

My pastor friend *was* an odd duck in the northwoods—a former boxer, a socialist, a dirt-poor working-class Australian by birth, a man alarmed by flag waving, the prison "industry," homophobia, and global warming. As he was packing his bags, I said something like this to him—Scott, take this idea to your seminary: tell them to prepare a handful of preachers, each honed on a specific subject. Somehow, through denominational meetings and ecclesiastical bureaucracy, get these issue-laden preachers into a rotating circuit.[1] Have them visit church after church, holding back nothing in their sermons for fear of their "careers." (I put "career" in quotes because a preacher is supposed to have a calling not a career and also supposed to believe that truth needs no clipping, trimming, or apologies.)

I know way too many preachers from a smorgasbord of denominations who lamely rationalize their pulpit fluff by saying their congregations aren't ready for the Gospel. I decline to comment on such a worthless thought. But the issue-laden, fiery circuit preachers might serve to wake everybody up. Or do we really prefer to doze our way to the funeral home, dressed in our Sunday best, our baptismal certificate securely tucked under the pillow, with the undertaker's best rouge on our cheeks? Maybe tuck in a note of explanation about how your broken bone got set.

---

1. Since writing this essay, I've been delightfully startled by the September/October 2002 issue of *Christian Social Action*, a magazine of the General Board of Church and Society of the United Methodist Church. On page 36 is a full-page listing of fifteen general areas (Climate Change, Death Penalty, Reproductive Rights, HIV/AIDS, Health Care, Rural & Urban Policies, Militarism & Disarmament, etc.) and twenty-three persons, complete with phone numbers and e-mail addresses, who are ready and willing to "preach, teach, and facilitate workshops, retreats, Sunday School classes, and forums . . . on timely issues of Christian social concern." Call Neal Christie at 202-488-5611 or nchristie@umc-gbcs.org for more information. (I called Neal Christie in early December of 2011 and, although the budget's been cut, the program's still up and running.)

26

An Admonition to Gore Vidal

THESE ESSAYS ARE INSPIRED, I suppose, in a manner not unlike how a landscape painter is captivated by a scene. They may be to some degree repetitive, but every essay paints a slightly different perspective. Hence their arrangement in an erratic essayistic gallery is not chronological but, in an improvised sort of way, thematic.

I have just now finished Gore Vidal's *Perpetual War for Perpetual Peace*. Like any other interesting set of essays, this one wobbles and wanders—the why and wherefore of the September 11 plane hijackers, the why and wherefore of Timothy McVeigh and his bombing project in Oklahoma City, and various other corners peeked around and fences peered over—but its theme (the title comes from a phrase of Charles Beard's) is essentially contained in a long paragraph near the end of the last essay, "A Letter to Be Delivered":

> Fifty years ago, Harry Truman replaced the old republic with a national-security state whose sole purpose is to wage perpetual wars, hot, cold, and tepid. Exact date of replacement? February 27, 1947. Place: White House Cabinet Room. Cast: Truman, Undersecretary of State Dean Acheson, a handful of congressional leaders. Republican senator Arthur Vandenberg told Truman that he could have his militarized economy only *if* he first "scared the hell out of the American people" that the Russians were coming. Truman obliged. The perpetual war began. Representative government of, by, and for the people is now a faded memory. Only corporate America enjoys representation by the Congresses and presidents that it pays for in an arrangement where no one is entirely accountable because those who

> have bought the government also own the media. Now, with the revolt of the Praetorian Guard at the Pentagon, we are entering a new and dangerous phase. Although we regularly stigmatize other societies as rogue states, we ourselves have become the largest rogue state of all. We honor no treaties. We spurn international courts. We strike unilaterally wherever we choose. We give orders to the United Nations but do not pay our dues. We complain of terrorism, yet our empire is now the greatest terrorist of all. We bomb, invade, subvert other states. Although We the People of the United States are the sole source of legitimate authority in this land, we are no longer represented in Congress Assembled. Our Congress has been hijacked by corporate America and its enforcer, the imperial military machine. We the unrepresented People of the United States are as much victims of this militarized government as the Panamanians, Iraqis, or Somalians. We have allowed our institutions to be taken over in the name of a globalized American empire that is totally alien in concept to anything our founders had in mind. I suspect that it is far too late in the day for us to restore the republic we lost a half-century ago.[1]

In the next-to-last essay entitled "The New Theocrats," Vidal brings forward the eighteenth-century Neapolitan scholar Vico who,

> . . . working from Plato, established various organic phases in human society. First, Chaos. Then Theocracy. Then Aristocracy. Then Democracy—but as republics tend to become imperial and tyrannous, they collapse and we're back to Chaos and to its child Theocracy, and a new cycle. Currently, the United States is a mildly chaotic imperial republic headed for the exit, no bad thing unless there is a serious outbreak of Chaos, in which case a new age of religion will be upon us. Anyone who ever cared for our old Republic, no matter how flawed it always was with religious exuberance, cannot *not* prefer Chaos to the harsh rule of Theocrats. Today, one sees them at their savage worst in Israel and in certain Islamic countries, like Afghanistan, etc. Fortunately, thus far their social regimentation is still no match for the universal lust for consumer goods, that brave new world at the edge of democracy. As for Americans, we can still hold the fort against our very own praying mantises—for the most part, fundamentalist Christians abetted by a fierce, decadent capitalism in thrall to totalitarianism.[2]

An Admonition to Gore Vidal

Vidal is certainly provocative. But one book and author against which to cast Vidal's Vico is *The Coming Caesars* by Amaury de Riencourt. The thrust of de Riencourt's analysis is contained on the opening pages of his Introduction:

> It is the contention of this book that expanding democracy leads unintentionally to imperialism and that imperialism inevitably ends in destroying the republican institutions of earlier days; further, that the greater the social equality, the dimmer the prospects of liberty, and that as society becomes more equalitarian, it tends increasingly to concentrate absolute power in the hands of one single man. Caesarism is not dictatorship, not the result of one man's overriding ambition, not a brutal seizure of power through revolution. It is not based on a specific doctrine or philosophy. It is essentially pragmatic and untheoretical. It is a slow, often century-old, unconscious development that ends in a voluntary surrender of a free people escaping from freedom to one autocratic master.
>
> New concentrations of power during the past fifty years of world wars, revolutions, and crises, have made this threat of Caesarism increasingly evident. Political power in the Western world has become increasingly concentrated in the United States of America, and in the office of the President within America. The power and prestige of the President have grown with the growth of America and of democracy within America, with the multiplication of economic, political, and military emergencies, with the necessity of ruling what is virtually becoming an American empire—the universal state of a Western civilization at bay.
>
> Caesarism is therefore the logical outcome of a double current very much in evidence today: the growth of a world empire that cannot be ruled by republican institutions, and the gradual extension of mass democracy, which ends in the destruction of freedom and in the concentration of supreme power in the hands of one man. This is the ominous prospect facing the Western world in the second half of the twentieth century.[3]

That's what Amaury de Riencourt was saying in 1957. But it's easier to see how Vidal's Vico is wrong, how his grand "organic phases" not only rest on a false initial phase, but how that falseness skews the understanding of subsequent phases—it's easier to see that by noting how de Riencourt falsely fingers *democracy* as the seedbed of imperialism, how *equality* supposedly produces Caesar.

The Kingdom of God Is Green

I've already quoted hard-nosed facts regarding wealth concentration from Kevin Phillips, Ralph Nader, and Frances Moore Lappé —how aristocracy has never left us, how feeble and puny have been the democratic inroads into aristocratic prerogative both economically and politically, not to speak theologically. (Liturgically speaking, God remains King of kings.) The wealthy own and control roughly as much wealth now, proportionately, as they ever did in civilization's bloody history. This is not an opinion. This *fact* of wealth concentration should at least cause us to squirm in our chairs a wee bit as we spout such pontifical nonsense as de Riencourt does in regard to the inherent totalitarian nature of democracy. (Though he was and is right about the growth of world empire and the concentration of power.) So, when empire becomes unmanageable—i.e., "cannot be ruled by republican institutions"—let's just blame its debility on an excess of *democracy*. That way, we are freed from the uncomfortable need to look closely or critically at "diseases" or "traumatic institutions."

Note that with Vidal's Vico the first "organic phase" is Chaos. (This is also Hobbes' first phase—each against all, dog eat dog—until Order comes in to clean house and straighten out the mess.) Please note that behind Toynbee's Civilization lies sluggardly, torpid Primitivity—even though Toynbee clearly recognized that Class and War, unless we undo them, will, because of technological lethality, kill us all. That is, since we fear an opaque Chaos composed of primitivity and backwardness, we will vigorously impose a murderous Order to relieve our anxiety. We resent a spirituality that requires vulnerability, and we worship the willful (and, if deemed necessary, violent) intelligence that rescues us from this threatening insecurity.

I've read a few of Vidal's historical novels over the years and, while he tells an interesting story, his bias toward sophisticated urbanity, toward "civilized values," can get a little overbearing. I say this because, to my surprise, Vidal not only takes an interest in Timothy McVeigh, Waco, and Ruby Ridge, he also (via Joel Dyer's *Harvest of Rage: Why Oklahoma City Is Only the Beginning*) spends the bulk of pages 61 through 65 on the plight of small-scale farmers, especially in the Midwest and Plains states:

> I read Dyer's *Harvest of Rage*. Dyer was editor of the *Boulder Weekly*. He writes on the crisis of rural America due to the decline of the family farm, which also coincided with the formation of various militias and religious cults, some dangerous, some merely sad.[4]

An Admonition to Gore Vidal

But Dyer has unearthed a genuine ongoing conspiracy that effects everyone in the United States. Currently, a handful of agro-conglomerates are working to drive America's remaining small farmers off their land by systematically paying them less for their produce than it costs to grow, thus forcing them to get loans from the conglomerates' banks, assume mortgages, and undergo foreclosures and the sale of land to corporate-controlled agribusiness. But is this really a conspiracy or just the Darwinian workings of an efficient marketplace? There is, for once, a smoking gun in the form of a blueprint describing how best to rid the nation of small farmers. Dyer writes: "In 1962, the Committee for Economic Development comprised approximately seventy-five of the nation's most powerful corporate executives. They represented not only the food industry but also oil and gas, insurance, investment and retail industries. Almost all groups that stood to gain from consolidation were represented on that committee. Their report [*An Adaptive Program for Agriculture*] outlined a plan to eliminate farmers and farms. It was detailed and well thought out." Simultaneously, "as early as 1964, congressmen were being told by industry giants like Pillsbury, Swift, General Foods, and Campbell Soup that the biggest problem in agriculture was too many farmers."...

So a conspiracy has been set in motion to replace the Jeffersonian ideal of a nation whose backbone was the independent farm family with a series of agribusiness monopolies where, Dyer writes, "only five to eight multinational companies have, for all intents and purposes, been the sole purchaser and transporters not only of the American grain supply but that of the entire world." By 1982, "these companies controlled 96 percent of U.S. wheat exports, 95 percent of U.S. corn exports," and so on through the busy aisles of chic Gristedes, homely Ralph's, sympathetic Piggly Wigglys.[5]

What is to be done? Only one thing will work, in Dyer's view: electoral finance reform. But those who benefit from the present system will never legislate themselves out of power. So towns and villages continue to decay between the Canadian and the Mexican borders, and the dispossessed rural population despairs or rages. Hence, the apocalyptic tone of a number of recent nonreligious works of journalism and analysis that currently record, with fascinated horror, the alienation of group after group within the United States.[6]

The Kingdom of God Is Green

Let's bring another author's name into this stew—A. V. Krebs, a Nader colleague. The book is *The Corporate Reapers: The Book of Agribusiness*. In Chapter twenty-seven, "Squeezing the 'Toothpaste,'" Krebs devotes several pages to the Committee for Economic Development, spelling out in greater detail the antiagrarian ideology of the high-powered corporate world, an ideology (if we can call it that) going back far beyond 1962. (If it's an ideology, it is doctrine derived from "founding violence" mythology. That is, as "ideology," the squeezing of small-scale agriculture is simply a continuation of civilization's impoundment of the early agrarian village, and that impoundment was an initial act of violence followed by the institutionalization of threat and acquiescence.)

I have dealt at some length with Krebs in a big manuscript that hasn't yet been published, entitled *This Populist Wraith: Ecological Socialism and the Resurrection of the Peasantry*. But I invoke Krebs here because his analysis, as trenchant as it is, is also too limited and short-sighted. Krebs essentially begins with Civil War-era industrialization in the United States, and he describes, chapter by chapter, decade after decade, how agriculture got squeezed as industry expanded. This analysis is true and good as far as it goes. But if fails to touch the ancient civilized backdrop to industrialization, the momentum of multimillennial, sustained exploitation by which the aristocracy lived off the peasantry. The aristocracy *did* live off the peasantry until technology enabled a new commercial aristocracy to liquidate the peasantry altogether and replace it with technologized agribusiness. What Joel Dyer chronicles are the death throes of the remnants of this ancient peasant culture—American-style, "farming as a way of life." Agribusiness is pure civilized agri"culture." (When technology destroys and replaces culture, is it really possible to call this new instrumentalist thing by the name "culture"? Civilization, strictly speaking, is much less an expression of culture than it is a peculiar sort of overbearing parasitism. The Greek *parasitos* means "one who eats at the table of another." So civilization-as-parasite is etymologically exact.)

I honestly cannot speak to the question of whether Islamic fundamentalism has a "concept" similar to the kingdom of God. I say this because the "Christian" militia movement Vidal pokes around in also seems devoid of "kingdom" understanding. They wallow, instead, in some weird sort of racial holiness mythology. All fundamentalisms seem disposed to invoke a very angry, very powerful, very righteous theistic God, a Mighty Father, who is gong to smash the infidels flatter than a pancake, with a little help from the Mighty Father's fervent sons.

The title of Vidal's next-to-last essay is "The New Theocrats," and in it he strikes out at religious superstition. He insists such superstition should be purged from secular politics. Well and good; but Vico's precivilized Chaos is also a superstition. (It's superstition because to imagine the precivilized world as Chaos is to totally misrepresent the organic stability of noncivilized cultures.) Chaos is a superstition that sits in the twilight between religious mythology and civilized mythology—between (one presumes) the remnants of ancient storytelling around the campfire (religious/folk mythology) and the self-serving ideological cover-ups and founding violence rationalizations of a triumphant ruling class (civilized mythology).

Of the two, truly naive religious/folk fundamentalism is far more honest and infinitely less corrupt. It may be wrong, and it may cling to its wrongness, but it's not inherently vicious. Religious fundamentalism is only a minor problem, except when its flames are fanned by those with a self-interested agenda and cynical motive: the way the Republican Party, say, fans the flames against those who "believe in" abortion. The real corruption, the actual cesspool of falsehoods, misrepresentations, mythologizing, and fabrication, lies in the camp of civility, among all those who, in their intellectual arrogance and spiritual immaturity, fear and loathe our common origin, are contemptuous of shared earthy simplicity, hate "backwardness," and strive mightily to achieve a perpetual "ambience of civility"—all the while denying or rationalizing the vicious predation on which civility reposes.

Civilization is the earthly manifestation or embodiment of a theistic Mighty Father. The state is His body. Congress is His left arm. The military is His right arm. The President is His head. That leaves the Courts either as His heart or—more likely as of late—His spleen. Embellishing on Thomas J. J. Altizer, we can say that when God died, civilization began to dress up in His dead body—a stolen garment it wears to this day. Perhaps this is not so much the Constantinian Arrangement as the Constantinian Wardrobe. Civilization proves that God is dead.

Notes

1. Vidal, *Perpetual*, 158–59.
2. Vidal, *Perpetual*, 143–44.
3. de Riencourt, *Coming*, 5–6.
4. Vidal, *Perpetual*, 119.
5. Vidal, *Perpetual*, 61–63.
6. Vidal, *Perpetual*, 64–65.

27

The Imposition of Those Glories

To REVIEW AND HOPE to do justice to *Constantine's Sword* is a daunting task. One could even review the reviews. (I've read five, the best by Charles Morris in the January 2001 *Atlantic Monthly*, the worst by Eugene Fisher in the March 5, 2001, issue of the Jesuit publication *America*.)

At one level, James Carroll's massive effort—616 pages of text, 68 pages of footnotes, a bibliography 24 pages long (even his Acknowledgements take two and a half pages!)—revolves around the question of whether, or to what extent, formal Christian teaching in regard to "the Jews" is responsible for the Nazi Holocaust or Shoah. Andrew Sullivan, in the January 14, 2001, issue of *The New York Times Book Review*, asks whether there is "a continuous link between this [Christian] Jew hatred and the final act of vengeance in the Holocaust?" He immediately answers his own question by saying that "Carroll is wise not to say yes."[1]

But Rabbi A. James Rudin, writing in the February 2, 2001, "Spring Books" section of the *National Catholic Reporter*, says Carroll "ruefully concludes that Christian theology itself is the root cause of 'Jew hatred': 'However pagan Nazism was, it drew its sustenance from groundwater poisoned by the church's most solemnly held ideology—its *theology*.'"[2]

Did James Carroll write two versions of *Constantine's Sword*? If so, I had the good fortune to read the same one as Rabbi Rudin—even as we will, for the time being, glide over the assertions that Nazism was "pagan" and that the sustenance for this paganism came from Christian theology.

If there is a book for Lent, a sustained meditation for Easter, this is it. (I write this on Maundy Thursday, the day of foot washing. It could hardly

The Imposition of Those Glories

be more appropriate.) If you have an interest in the powerfully complex meaning of the cross, whatever your religious persuasion (or lack of one), this is the book for you to read slowly and with great care.

But I want to be very clear: In my estimation—and I readily confess to a lack of qualification to make this outrageous statement—*Constantine's Sword* may well be one of the biggest threats to conventional Christian self-understanding ever published. Carroll's book calls on the church, especially the Roman Catholic Church, to repent of its ancient and pervasive anti-Semitism and to reform itself from within. But to repent of its anti-Semitism, the Church has to come to grips with the pivotal place "the Jews" have had in Christian theology for well over fifteen hundred years. This is nowhere as simple as it sounds. This is an analysis that precedes revolutionary change.

What Carroll is exposing and challenging is the conventional, traditional Christian understanding of power and, even more potently, the church's historic collusion with power. In his review, Andrew Sullivan says that "from Constantine's sword, designed in the shape of a cross, the fusion of a religion opposed to power with power itself is the core of the corruption of Christianity. When Christians used this secular power to persecute, banish, murder Jews, they were betraying not just the essence of the faith of Jesus, they were embodying the very power that killed Christ—not the evil Jews, but the power of the state."[3] And that is *exactly* the point.

This story of power does not, however, begin with Constantine's "conversion" in 312—that *political* act resulting not only in the church's forced marriage *to* Empire, but also in the church's new identity *as* Empire—even as the cross was transformed from crucifixion horror icon to triumphant religious weapon—i.e., "Constantine's sword" became a new image of the cross as a religious charm to ward off or conquer evil. This new image of, and new regard for, the cross changed the entire understanding of the meaning of the cross from a terrorizing instrument of brutal execution to an object of religious veneration and even worship. This transformation of the ethical meaning of the cross is symbolic (that is, it replicates in iconic form) the transformation of Christian self-understanding from kingdom of God conviction to empire consciousness. In his second chapter, Carroll says that "Lenny Bruce, the Jewish shockmeister, used to send a naughty thrill up the spines of his audiences by professing relief that Jesus wasn't born in twentieth-century America, because then, Bruce would blithely aver, pious Christians would have to wear tiny electric chairs around their necks. In fact, the cross did not serve as a Christian icon until it ceased being a Roman execution device in the fourth century."[4] So, is wearing or displaying the cross as "Christian icon" an unconscious acknowledgment of complicity in the Constantinian Arrangement?

In some ways, Carroll's analysis boils down to a painfully simple question. That question is: Who killed Jesus? Was it "the Jews"? Was it God? Was it civilization in its specific Roman manifestation?

I'm going to bring out my double-bitted axe here, one I've been grinding on—and I hope sharpening—for years. I have two very determined interests. One is, for all my hesitations, reluctances, and denials, to know more fully who and what Jesus is, what his teachings demand of me, and what the "kingdom of God" really means. I say this because we humans need foundational ethics and spiritual grounding. Others may find that grounding in their own way. A major portion of my grounding, such as it may be, came as Gospel stories simply told and deeply felt. My other interest is hugely related: how civilization has sucked the life out of agrarian culture and thrown away the rustic corpse.

To understand much of anything of Jesus' peasant parables, it seems to me, one also needs to grasp the crucifixion—I do not use the word lightly—of peasant culture by the traumatic institutions and diseases of civilized power. Just as Roman civilization killed Jesus (we shall soon see that Carroll explicitly affirms this thought), so has civilization in its industrial globalization exterminated virtually all forms of noncivilized culture. If I have a bone to pick with James Carroll, it's that he ends up shielding the atrocities of ancient Rome and Nazi Germany by designating these atrocities "savage" and the political organizations "pagan." By this oxymoronic slight of hand, a semantic absurdity and conceptual magic fully accepted in the professional shuck and jive of historical analysis, civilization emerges clean and bright and "pagan" gets nailed.

Now "pagan" shares a common Latin root (*pagus*) with "peasant," just as "savage" derives from a Latin word (*silvaticus*) for woods or forest. So what is most *removed* from civilized power gets blamed, implicitly if not explicitly, for the systemic cruelties of civilization. There is a name for this process. It's called scapegoating, and scapegoating is largely a method of evading difficult or painful truth, of pinning the Devil's tail on somebody else's donkey. It's the core of what Gil Bailie understands as myth. Despite his painfully honest exploration of how Christianity was, by Constantine, forcibly wedded to Empire, and how Christianity built its theology on the backs of scapegoated Jews, Carroll declines to untie this particular conjugal knot of illicit marriage and instead scapegoats "pagan"

The Imposition of Those Glories

when it comes to ultimate blame. It is a totally conventional, totally acceptable, and totally outrageous false analysis.

The ultimate atrocity against Jesus and the Jews was not perpetrated by "paganism." The executioners of Jesus and the Jews were direct agents of church and state, the essential core of Western *civility*.

The core problem with Carroll's book is the core problem with Christianity: how "civilized" each is. We have almost no idea how totally we've idolatrized civilization by making it the faultless object of our secular worship. It is, after all, the real husband of the so-called Bride of Christ. Or, to put it differently, if the early church was the Bride of Christ, then Constantine as emperor claimed *jus primae noctis*, the right of a feudal lord to the virginity of a female vassal on her wedding night. Do I wildly overstate? I repeat my basic comparison: if civilization killed Jesus because he represented everything civilization despised, hated, and feared, then civilization also killed agrarian culture because sufficiency, simplicity, sharing, living respectfully close to Creation (nature, if you prefer), and distrust of arrogant organizational perfection all stood in the way of maximizing civilizational utopianism. The crucifixion of Jesus prefigures the crucifixion of the peasantry as a whole, just as the "marriage" of Christianity to Empire prefigures the forced corruption of agriculture into industrial agribusiness.[1]

It's time to let James Carroll speak for himself.

Carroll's ambivalence—shall we call it a "civilized" ambivalence?—shows itself most clearly in a chapter ("The Church and Power") at the end of the book. He says the "Church has never come fully to terms with the contradiction it embraced when the Roman imperium and Roman Catholicism became the same thing."[5] But then he quickly says this:

> The point is not to wish sentimentally that the Christian religion, in order to maintain its purity, had remained the marginal cult of a despised minority, with an ad hoc organization; more charismatic than catholic; possessing nothing; being more acted upon than a spur to action; and innocent, not in the sense of sinless but in the sense of untested, untouched.[6]

I will go on with this quotation—one long paragraph. But first it's necessary to point out how Carroll frames the alternative, what language he employs: the "wish" is sentimental, in order to maintain "purity," a

---

[1]. See "Redemption of the Past," Chapter 10 in my *Nature's Unruly Mob*, especially pages 98 and 99, for a glance into the metaphorical and metamorphical implications of Jesus' crucifixion.

"marginal cult" of a "despised" minority, with no real organization, having nothing. But, oddly enough, he also employs a strange combination of "being more acted upon" than acting while also (what could this possibly mean in light of the history of brutal martyrdom?) "innocent" in the "sense of untested, untouched." That is to say, Carroll describes the alternative to the imperial church as a wishful weakling, and he brushes it aside.

But it seems to me he's not only brushing aside a straw man—or a straw church—he's also beginning to scapegoat this early, precivilized church with its alleged weaknesses and untestedness in order to plump up and justify the post-Constantinian church. Here's more of that paragraph:

> The glory of the Church includes what its institutionalization has enabled—a transcending of time and culture, a triumph over history that stands alone. Every Catholic is proud to be part of a two-thousand-year-old tradition that still lives.[7]

Note that the preimperial wishful church was, by implication, no institution, that it (also by implication) was not able to "transcend" time and culture, was not a "triumph over history that stands alone." Oh my goodness! How heroic! And Carroll's book on the "triumph over history" opens with the Triumphant Cross of the Triumphant Church planted on the Triumphant Grounds of Triumphant Auschwitz! "Every Catholic is proud...." These sentences are pure, unadulterated, repulsive salesmanship. Every Catholic—every Christian— should be ashamed. Back to Carroll:

> But such longevity presumes a weighty bureaucracy. It presumes possession, even wealth. It presumes the great contests of will between rulers and popes, princes and bishops, masters and monks.[8]

Does it presume the ownership, in medieval times, of a vast portion of Europe, of huge estates? Does it presume the continuation of civilization's economic suction pump in every peasant village, the maintenance of a devious ideology of economic extraction, disguised as the suppression of "paganism," over which it rules supreme?

> It presumes the wily strategies that survival required. It presumes the yoking of intellect to piety, and the adaption of faith to Plato's separation of form and matter, to Aristotle's rational quest for universal order, and ultimately to Kirkegaard's leap into the arbitrary. It even presumes the admission of politics to pulpits, and perhaps the conscription of cloisters into service as bastions. The ruins of Europe, the museums of Europe,

The Imposition of Those Glories

the cathedrals and castles and Gothic-towered universities of Europe, are the stone record of this story, and though there is cold shame in it, there is carved beauty too. The history of the Church, not above the world but in it, only continues what we just saw revealed by the Church's troubling foundational texts—that the Church is of the human condition, not against it.[9]

In this one long paragraph we have it all: the alternative to the imperial church is a sentimental wish; the glory of the institutional church is an enabling transcendence over time and culture, a triumph over history, as it slugs it out with rulers, princes, and masters; and its wily strategies of survival, including manor lordism, only continue the reality that the church is "of the human condition, not against it." This is everything and nothing. This is all things to all people, and everybody gets to go home cold and hungry. Carroll's book, thank God, is infinitely greater than this miserable paragraph. But in these rotten sentences, Carroll exhibits an ambivalence that (one suspects) is at work in him always: nowhere, but for a fine passage I'll shortly quote, does he even talk about the life condition of "peasant peoples," of what all this imperial glory might have meant for *them*.

In his massive investigation into the history of the church and the Jews, Carroll shows that the Jewish people have been the target of an antagonistic Christianity since the first century. What sort of target? At first, the new Jewish "Christians"—devoted followers of Jesus—had a family fight with skeptical Jews over this new spiritual articulation. As non-Jews became numerically dominant in what was becoming a new religion, and as Rome finally crushed Israel and destroyed the Temple in Jerusalem, Christian writers began to overtly blame "the Jews" (as in the Gospel of John) for the crucifixion of Jesus.[II] In Chapter 9, "Jesus a Jew?"—Carroll says:

> There is perhaps something craven in the Gospels' emphasis on "Jews" as a threat to order in the empire, as opposed to "Christians," and it does not mitigate the Gospel writers' responsibility for driving this wedge to note that they were responding to Roman oppression. But the more fundamental point is that in doing this, the followers of the murdered Jesus were only demonstrating how effective the imperial overlord had been

---

II. The term "the Jews" occurs nearly sixty times in John's Gospel, nearly all those usages couched in a pejorative or disparaging ambience, including in scenes (e.g., John 18 and 19) purporting to show the innocence of Pilate, the top Roman in charge. Scapegoating, in other words, is already built into the fourth Gospel.

The Kingdom of God Is Green

> in infecting the dominated population with its own cynicism and contempt. This dynamic becomes even clearer in the context that has provided us our starting point: One measure of the diabolical efficacy of Nazi torment in Auschwitz, besides the way Jews were victims of SS guards, was the way Jews were victims of fellow Jews, the capos who served as SS surrogates. The collapse of the moral universe that led Jews to participate in their own destruction in the death camps, or to take upon themselves a feeling of guilty responsibility for the evil around them, only emphasizes the abject evil of an absolutely oppressive system. That evil lies in the system's capacity to destroy the innocence of everyone it touches. When Jewish factions turned Rome's venom against each other, Rome won yet another victory. There is no question here of "Christian innocence," because among human beings there is no innocence when the question becomes survival. Extreme violence and extreme measures to survive it form the ground on which this entire story stands.[10]

In the next chapter, "The Threshold Stone," Carroll goes on to say that one of the things

> ... wrong with blaming the anti-Jewish texts of the New Testament on a primitive and essential "Christian" hatred of Jews is that doing so continues the victim's habit of exonerating the true villain in the story, which was and remains Rome. I acknowledge the apparent absurdity of this attempt, two thousand years after the event, to reconstruct its shape and meaning with more accuracy than the people who lived only a generation or two later. But in this one regard at least—the crucial influence of a dominant overlord—we have a distinct advantage over those first Christians and rabbinic Jews. For us, the grip of the overlord has long since been released, and the myth of hierarchy has been broken. The blanketing fog of an imperialist occupation blinded those who lived through it to the all-encompassing nature of Roman oppression. Similarly, the Romans, by controlling the future, controlled the way even their extreme savagery would be remembered by Jew and non-Jew alike. Yet neither of these facts excuses us from emphasizing that the story of Jesus, at a fundamental level, is one part of the story of Israel's refusal to yield to Rome. And this can be perceived more clearly now than it was then.[11]

The Imposition of Those Glories

The "myth of hierarchy" now broken is, of course, ancient Rome's—and only ancient Rome's. But, again in Chapter 9, Carroll provides us with a view "from below":

> It is the glories of Roman dominance that are emphasized in the cultural memory of Western civilization—those arrow-straight roads, elegant aqueducts, timeless laws, conjugated language—to the exclusion of what the imposition of those glories cost those on whom they were imposed. What if, when we thought of Caesar, we thought less of Cleopatra's lover or Virgil's patron or Marcus Aurelius's delicate conscience than, say, of a Joseph Stalin or a Pol Pot whose program worked? How would history tell the story of the twentieth century if it were the first century of the thousand-year Reich? It all depends on where you stand. It may be anachronistic to judge the policies of a great empire of antiquity by the standards of the U.N. Declaration of Human Rights, but if being human means anything, it is that a minimal level of decent treatment is required in every culture and era. It is clear that from the point of view of those on the bottom of the Roman pyramid—indeed, under it—that such a minimal standard was not met.
>
> To the peasant peoples of the Roman-dominated world, to the millions of slaves and petty laborers (in Rome itself, fully one million of the population of two million were slaves), to the lepers and beggars, to the troublemakers whose lives could be snuffed out with little notice taken, no characterization of Caesar's evil would have been too extreme. We have looked back at Rome from above—from the point of view, that is, of those who benefited from its systems, traveled its roads, beheld its architectural wonders, learned to think in its language—but what of that vast majority who drew no such benefit? There is no understanding either the Jesus movement itself or the foundational memory of its violent conflict with the Jews if we cannot look back from below, from the vantage of those for whom the Roman systems were an endless, ever-present horror. It was to them, above all, that the message of Jesus came to seem addressed.[12]

So, a view from above enables us to glorify hierarchy—literally the "rank or order of holy beings," according to *Webster's*—while exonerating the "true villain." ("Villain" is one of those Trojan horses with a slovenly scapegoat tucked etymologically and embryonically inside, waiting to spring out and track barnyard filth across the clean floors of our elegant, timeless civility. Villain comes from *villa*, farm or country house, is of course closely related to "village," and means a "baseborn or clownish person; a boor"—that is, a farmer or a peasant. Why *are* so many of our cruelest epithets directed at the countryside and at the people who live and work there? Could this have anything to do with how obsessively civility hides its founding violence by endlessly scapegoating its victim?) And a view from below (as in Luke 22:37, where Jesus desires to "let himself be taken for a criminal"), contrary to all of Carroll's gushiness "from above" about the imperial Church's "triumph over history," is precisely to witness hierarchy—including, most emphatically, ecclesiastical hierarchy—as (full semantic circle here) the "true villain."

Why can't we just say that the peculiar political organization that came into being by preying on and expropriating the abundance of agrarian villages in the late Stone Age is based on selfishness, greed, scapegoating, and economic predation? Why can't we say that *civilization*, with its aristocracy, hired swords and coerced slaves, is a world-class criminal? Why can't we say this? Is psychological deference to the imperial overlord so deeply ingrained we can no longer do without it? Has civilized mythology so deeply taken root in us that we are reflexively horrified at the prospect of being or becoming uncivilized? What happens when Constantine grabs Jesus by the throat, jams him back on the iconic Holy Cross, and makes him the Mascot Prince of Empire? Suddenly, he is no longer a revolutionary savior in behalf of radical servanthood and radical stewardship but instead a Fascist Prince who demands obedience to imperial authority. Amazing. Truly and absolutely amazing.

Somehow, over time, crossing and recrossing the line between being overtly victimized by hierarchical program and propaganda, and being uncertain how to live or act without the habitual directives of hierarchical propaganda and aristocratic program, we fall prey to the mandatory worship of hierarchy, civilization, progress, and perfection—all of which locks us, as it locked our ancestors, into an ongoing psychocultural dependency from which we seem incapable of extracting ourselves. And so the wheel of sacred civility keeps turning, crushing, and poisoning. (Do we need other reasons to explain our ethical oblivion in regard to nuclear weapons or climate change?)

Carroll follows the church's policy toward Jews through the centuries. While only on occasion openly murderous, that policy was always, except for rare persons like Abelard or Nicolaus of Cusa, hateful and contemptuous: Jews as "Christ killers," Jews as a kind of spiritual sickness within the Christian social body—much as "paganism" is a spiritual disease Christian civility has felt duty bound to extirpate. Carroll (himself a former Catholic priest) recognizes the overt linkage, from Constantine onward, between empire and an enforced doctrinal "purity." Yet, without so much as a hint of hesitation, Carroll repeatedly employs the loaded term "pagan" in its totally conventional usage. The old Roman gods were "pagan." Roman heroes were "pagan." Nazi ideology was "pagan" or "neo-pagan." In a work whose very purpose is to tunnel into and expose the church's scapegoating of "the Jews," Carroll blithely maintains the scapegoating of "pagan"—as if the term was self-explanatory to the (presumably) educated reader.

The tragedy, of course, is that "pagan" *is* self-explanatory as a scapegoating buzzword whose hidden purpose is always to deflect attention from the colossal crimes and inherent criminality of civilization—and also, in orthodox Christian terms, to—not so much deflect—terrify people in regard to the deep and pervasive evil lurking in the *pagus*. If Judaism was traditionally identified as Christianity's "negative other," then paganism (and a small but powerful set of similar words with wilderness, rural, and agrarian roots) is civilization's "negative other." Civilization's foundational crime was—and is—the murderous armed robbery of the agrarian village, and so civilization's choice epithets contain denial of this crime in the form of metaphysical accusations against the victim. Christianity, by becoming civilized, also turned its identification of wickedness toward the countryside and those who live there.

Carroll can also say, in Chapter 11, that "What Jesus spoke of, and in his life embodied, was the opposite of Roman domination."[13] Yet Carroll has no name for this "opposite." The term "kingdom of God" hardly appears in his book; but where it does appear, its use is explicitly in reference to the "Second Coming of Jesus."[14] That the "kingdom of God" might mean Jesus' teachings manifest as radically alternative social behavior and a radically different political policy may be (one could argue) implicit in Carroll's work, but it is certainly not explicit.

Perhaps I wear my generic Protestantism too openly on my sleeve, or a peasant chip too cockily on my shoulder, but I believe I detect in James Carroll a deep reluctance to ask what civilization is or to frankly assert that the "kingdom of God" fundamentally confronts and contradicts civilized predation. He certainly doesn't ask what "pagan" is or has come to represent in conventional analysis. And to those who say: Words change over time and take on new meanings, I reply: Yes, they do; but the very process of change can itself be the result of scapegoating evasion—and that is exactly the function "pagan" plays in protecting civilization from ethical scrutiny.

Yet Carroll calls Rome the "base villain in the story," an evil whose systems were an "endless, ever-present horror." And he demonstrates conclusively—no surprise—how the church wedded and began to merge with empire on the occasion of Constantine's "conversion." My point is this: We will never adequately grasp the deeper implications of Christianity's scapegoating of the Jews until we grasp the larger, more generalized and diffuse scapegoating of peasant and pagan by civilized imperialism, for that is how we begin to grasp (or to be grasped by) the "opposite of Roman domination" that "Jesus spoke of, and in his life embodied."

Carroll does not say that Christianity's long hostility toward the Jews made Hitler be who he was; but he does say that Hitler cannot be understood, the roots of the Holocaust cannot be adequately grasped, without recognizing the central, pivotal role the civilized church played for centuries in fostering and institutionalizing anti-Semitism. Anti-Semitism and contempt for the pagan are scapegoating cousins. Both serve to protect the founding violence of civilization. Carroll, in the end, does confront, with painful honesty, not only Roman Catholicism but all of Christianity, and he calls for deep personal and institutional repentance—for the church's anti-Semitism, and for Christianity's theological imperialism. He calls for major reform of the institutional church.

But I don't believe that Carroll grasps the sheer magnitude of change implied by his own analysis. When—if—we stop playing a shell game with scapegoats, we will face a total reorganization of human life on Mother Earth. And then we will really look to people like Jesus, Siddhartha, and Mohandas Gandhi for guidance in servanthood, stewardship, and village-mindedness. Perhaps we might even better understand what Dietrich Bonhoeffer meant by "religionless Christianity." But, until then, "repentance" remains a little weak—or is it stiff?—in the civilized knees.

Notes

1. Sullivan, "Christianity's," 5.
2. Rudin, "Centuries," 30.
3. Sullivan, "Christianity's," 5.
4. Carroll, *Constantine's*, 15–16.
5. Carroll, *Constantine's*, 570.

6. Carroll, *Constantine's*, 571.
7. Carroll, *Constantine's*, 571.
8. Carroll, *Constantine's*, 571.
9. Carroll, *Constantine's*, 571.
10. Carroll, *Constantine's*, 87.
11. Carroll, *Constantine's*, 89.
12. Carroll, *Constantine's*, 80–81.
13. Carroll, *Constantine's*, 117.
14. Carroll, *Constantine's*, 567.

28

The Village of God

JOHN SHELBY SPONG IS a retired Episcopal Bishop of Newark, and in *A New Christianity for a New World* he says that his "admired friend and fellow priest Matthew Fox suggested groundwater as an analogy by which we might think of God in this post-theistic world. I find that image quite compelling."[1]

I'm not smart enough to know where one draws the line on what constitutes an "image." If metaphor, as *Webster's* says, can also be "regarded as compressed simile," and if metaphor is a way of suggesting likeness or analogy, then to say God is like this or that gets us chasing the tail of images pretty fast. There may be almost no way out of this doorless and windowless room (*Webster's* little SuperMax of the imagination) except by focused spiritual concentration (as, perhaps, via the discipline of Buddhist meditation) which then enables one to sink right through the floor . . . into groundwater. (Is it watering the image too much to invoke Rabbi Rudin's language, from his review of *Constantine's Sword*, of groundwater *poisoned* by "the church's most solemnly held image—its *theology*"?)

Images are tricky—an image of God is not an experience of God, and a mere image can even serve to deflect experience—and yet we humans are always describing things and depicting scenes to one another, giving directions, explaining how good the food was, how savory the odors were, how glorious the sunset, how tender the kiss. To say we get nothing out of description is laughable; we communicate with and by images all the time.

This is also true of our spiritual life. But here, perhaps especially for those of us raised in one of the three Abrahamic religions, we run into a prohibition against images. An "iconoclast" is, literally, a breaker of

The Village of God

images. The God of Moses was so jealous of other gods that He forbade the making of images (Exodus 20:4), which is an injunction full of divine whiplash.

Whiplash, of course, is to be struck by a whip, to be punished for an infraction of the rules by a jealous God who really lays it on for violation of His law. This God is powerful, He is male, He is a Person who thunders, exudes smoke, and has divine biceps.

Whiplash, of course, is what can happen when you've been rear-ended by a speeding concept. It jerks your head around and may even necessitate spiritual traction. A jealous God may turn out, after all, not to be a psychotic, self-obsessed Male in a constant control frenzy over threats, real or imagined, to His spiritual harem, but, instead, an unimaginably loving God (or Goddess) who so desires us not to wander off after mirage gods, into dead-end, desolate canyons, that She, as Shepherdess, will tuck ninety-nine sleepy sheep in for the night and go looking for the stubborn little lamby with the unconventional attitude who's gone in search of more truthful (or more lushly Green) pastures.

These are still anthropomorphic images—God as fierce, butt-kicking Chieftain; Goddess as tender, self-sacrificial Shepherdess. Although, in my opinion, we could use—we desperately need—a lot less of the former and a whole lot more of the latter, there's also *groundwater*. That is, we need to get loose of the image of God or Goddess looking human or superhuman. (Is theism tightly linked to an image of the divine as Person? Can God be Creator and Ruler without also being a Person? If God can be a Person, why can't a corporation?) But that we might think of God as groundwater, in a time when groundwater is, literally, being poisoned by civilized chemicals, and when the peasantry is, also literally, in the process of being globally extinguished by civilized economics, implies that traditional Christian theology has huge contributed to this poisoning, especially with its contempt for the natural and for the countryside—that is, for the *pagus*.

Perhaps that's why John Shelby Spong devotes so much of his *New Christianity* to the hammering of theism. Perhaps theism is philosophical hypothesis posing as religious conviction.

Theism has puzzled me for a long time. It's invariably struck me as odd that the red carpet leading to the sanctuary is philosophical. That is, we first believe a thing to be true, then we construct a religion we say is, in some obscure way, rooted in experience. That is to say, philosophical assertion (i.e., God exists or God created the universe) is the text, all the

The Kingdom of God Is Green

rest is gloss. It's a book with as few as two words in it; but the religious footnote is voluminous.

My hunch is that John Spong's hammering of theism comes out of his intense frustration over religion being philosophy's resplendent footnote. For religion (perhaps the better word is "spirituality") to really be more real to us, we have got to get past, in our spiritual murkiness, merely *believing* in God as a pious mental exercise and come to a place where we actually *experience* Spirit as the Groundwater of Being. Either the real eutopian thing or give up the utopian fantasy.

In his hammering, John Spong whacks six-day creation, the parting of the Red Sea, the sun standing still so the ancient Hebrews can win their battle, the virgin birth, the literal resurrection of Jesus. Spong's a bit of a burly blacksmith with a big hammer in a delicate and divine china shop. All this is not terribly exciting (the china pieces have been whacked so many times by so many people they're getting to the china dust phase); but beyond all this whacking John Spong has a mission, a purpose, an insight, and an understanding. That insight is what he calls the "realm" of God.

John Spong intuits that Jesus really meant something by all the "kingdom of heaven" and "kingdom of God" talk. He intuits that the yeast metaphor in the parables has historical morphology and biological traction. He senses that, whatever else is in the spiritual mix, Jesus really had in mind a transformed social order, and that the very real threat this new social order posed to the existing state (both Jewish and Roman) is what triggered his (legal) assassination. Spong also recognizes that the concentrated energy of Jesus, including the vibrant dynamism of his "kingdom" teachings, was weakened, dissipated, and largely lost as this new religion (Christianity) became increasingly abstract, authoritarian, centralized, orthodox, devoted to otherworldly salvation, and governed by creedal assertions that asserted the purity of right metaphysical belief at the expense (a very expensive expense) of living compassion unfolding in ecological community.

John Spong no longer wants to find a dead Jesus at the end of a philosophical carpet embroidered with explicit crucifixion creeds. He wants to discover a living Jesus fully engaged in creating the "kingdom of heaven" in the everyday here and now.

But the language causes problems. For John Spong, theism is the core organizing principle of human behavior since the dawn of human consciousness. This is, for me, a new thought, and it's one I'm not exactly prepared to comment on—except to say that, if it's true, then civilization is a peculiar case—an astonishing intensification—of theistic unfolding that merits close and special attention. John Spong does not, however, render civilization this special attention. And although he clearly recognizes patriarchy and pyramidal hierarchy as two attributes connected to civilized theism (God is "King of kings"), and he therefore chooses to substitute the "realm" of God for the "kingdom" of God, he doesn't acknowledge that it's kings who have realms. He doesn't explore the mythic terrain of founding violence by which kings come into possession of their realms.

In other words, despite the recognition of semantic deficiency in traditional "kingdom" language (John Dominic Crossan also wrestles with this problem on page 266 of *The Historical Jesus*), "realm" doesn't cut the mustard either. It only kicks the can down the road. If the language is to mean something pervasive, growing, spreading, and transforming—something even digestively biological—then the "kingdom" of God is a cultural, social, and psychological petri dish, a virus, yeast that is slowly and steadily infecting the entire human race. If civilization is a power complex that, by virtue of its unbridled aggressiveness has globalized the world for reasons largely having to do with predatory self-interest, then the question becomes: What happens when the "kingdom" or "realm" of God infects and leavens civilization? Does it mean gentler and kinder bombers? ICBMs with blessed warheads? Capitalism with a Jesus face? Corporate executives consulting their WWJD wristbands? Spiritual technofix and poisoned groundwater control teams?

The preoccupation of my adult life has been the plight of small-scale agriculture and the crucifixion of the peasantry as civilization in an industrial mode found a new way (agribusiness) to produce food in huge quantities without peasants, without farmers. Civilization, in its industrial mode, has utopianized agriculture by forcing massive quantities of food from chemicals and terminator seeds. Industrial civilization industrializes agriculture, much as ancient Rome created huge estates whose moving parts were slaves. One can get into arguments about whether Jesus was a peasant or whether parables are a kind of peasant talk, but it's very clear—to me, at least—that the "kingdom of God" (whatever else it might be) is a close-to-the-Earth social order liberated from civilized oppression and utopian exploitation, the antithesis of "lording it over," a community all through, and a largely subsistent culture abundantly rich in stewardship and servanthood.

Therefore I invoke Gandhi's village-mindedness and Wendell Berry's longing for ecological settlement in undefiled landscape. Therefore I propose not the "kingdom" or "realm" of God but the *village of God* or, alternatively (thinking of Neil Douglas-Klotz's *malkuta*), the *village of the Goddess*, as a term that not only points us in the right spiritual direction but also resolves the unsatisfactory king/kingdom/realm business. (Marcus Borg, at the end of his delightful and stimulating *Meeting Jesus Again for the First Time*, talks about the "alternative community of Jesus."[2] Although this gets us out of the "realm" problem, the term "alternative community" simply doesn't have adequate semantic gumption to convey overflowing ecological abundance ethically shared.)

If theism and civilization are closely linked (theocracy is their fusion), and if civilization is a special case of theism (or, at minimum, if civilization uses theism to sanctify its traumatic institutions), then the end of theism spells the end of civilization as we know it. This is not so much the "end" of God or the "death" of God as it is the recovery of the Mother, the humbling of the Father, the liberation of the Son, and the radiant unfolding of the Daughter. Such a notion might amuse, aggravate, or terrify us. That it can do any of those things is caused by, rooted in, our overwhelming identification with civilization and the Christian Trinity—with their value, purpose, and sanctity. Civilization is, as I have elsewhere asserted, our actual religion. It's really what we cling to and believe in. Therefore the yeasty overtaking of civilization by *malkuta* is the transformation of predation by love, of utopian purity by earthy compassion; and, under the influence of love, bombers cannot bomb, greed shrinks, stealing ceases, and a new ecological social order emerges worldwide, based on stewardship and sharing.

This new social order, in the absence of predation, will begin to more generally resemble a healthy, wholesome, self-regulating village than a kingdom or realm with all its elegant, elite instruments for enforcing control, maintaining expropriation, and selling ersatz aristocratic products to a bewildered population beaten free of its self-provisioning capacities and folk consciousness. In this "village" of God, respect for human life will be intense and reverence for Creation will be fierce. Life will be abundantly frugal and ecologically clean.

And that brings us back to groundwater.

If God (in Paul Tillich's term) is the Ground of Being, then it's not a huge stretch to ponder groundwater as "an analogy by which we might think of God in this post-theistic world"—this image John Shelby Spong says he picked up from Matthew Fox.[1] The thing is, groundwater has

---

1. As I edit these essays prior to publication, I realize with increasing clarity that

become increasingly contaminated precisely as civilization has extinguished the peasantry and imposed chemicalized agribusiness. Utopia is toxic. Pure doctrinal intelligence is ecocidal madness—risen so far above both nature and folk consciousness that it imagines itself perfect and infallible. The peasantry has been crucified by civilized power and was, in my estimation, the bulk of the dough—the whole wheat part—the kingdom of God worked to leaven most easily and compatibly into something—not perfect—whole and wholesome. To be consistent within the metamorphosis of Christian theology, the crucifixion of the peasantry requires its resurrection.

Whatever the actual meaning of "resurrection" and "ascension"—not as general abstract concepts or mere mental images—we know, because it's two thousand years' worth of history, that something of Jesus lives on. Call it the Holy Spirit if that's what you prefer. I'm not in the mood to quibble over words. Call it—Jesus perhaps did, if we are to believe Neil Douglas-Klotz—*malkuta*.

Groundwater, yeast, the tiniest of seeds—all these are images that enable us to glimpse (get a mental picture of) how the spiritual really works. And something wonderfully mysterious happens—we slowly become cognizant of it occurring—as we not only *think* certain thoughts interesting and attractive, but we literally begin to *feel* a spiritual unfolding within our conscious bodies: no longer a grim and dogged determination to merely *believe* a grocery list of requisite ideas and creedal formulations, but, rather, a bodily *knowing* that love operates on the wave length of glorious eternity and that hate is a black hole going grimly nowhere.

Therefore the seed forever sprouts, the yeast leavens the sodden dough, and groundwater wells up into gurgling everlasting life. The "king" comes round the table with a tub of warm water and a clean towel to wash and dry our dirty feet. The "kingdom" is lush with gardens, orchards, fields, footpaths, ponds, and snug cabins permeated with the odor of freshly baked bread. There is singing in the air.

To drink deeply of this groundwater is to be slowly and steadily purged of greed, anger, violence, retribution, raw lust, and pride. To be thus purged is to both be content in Creation and to reverence the Creator: to live in such a way that groundwater is no longer poisoned. This

---

"God" for me has become a designation indicative of theism, while "Spirit" (to use Spong's term) is both pre- and post-theistic. Spirit undergirds or infiltrates all ages. If the Mother was grounded on Earth, the Father had us all looking up. And if the Son lived on Earth but ascended, perhaps the Daughter is breaking free of her Electra complex and is in process of descent.

The Kingdom of God Is Green

living is both intensely personal and communal; the village is the nucleus of ecological community. The twilight of exploitation is the dawn of sharing. The end of predation is the beginning of stewardship. To live in this resurrected village is to witness the resurrection of the peasantry, and the resurrection of the peasantry leads directly to the village of God. The kingdom of God is Green. The *malkuta* of the Goddess is gorgeously Green and wonderfully vibrant.

Notes

1. Spong, *New*, 183.
2. Borg, *Meeting*, 135.

29

This Predator Beast Game

BARBARA EHRENREICH, IN BLOOD *Rites: Origins and History of the Passions of War*, takes us on a very compelling tour. She drops us (and herself; she's the guide) behind the lines of civilization into the murky and opaque uncertainty of "pre"history. She takes us back to a time when human beings—before campfires, before bows and arrows—were as much prey as predator, when bigger and stronger beasts might drag a human home for lunch. She wants us to consider whether or to what extent a certain deep dread was built into us at that time, a dread never really exorcised in subsequent social developments, a dread still operative in human consciousness and in human institutions, a dread of predator beasts.

To summarize *Blood Rites*, or to try and lift a compact sampling of quotation (such as I have done with John Dominic Crossan and Thomas J. J. Altizer), is a rather dismaying task—dismaying because the brilliant insights and provocative assertions in *Blood Rites* are so densely packed. But here goes:

> Our understanding of war . . . is about as confused and unformed as theories of disease were roughly 200 years ago.[1]

> There is something in us, or at least something in some of us, that urgently seeks to make sense out of disconnected data and unassimilated experience, to draw links between people "like us" and people not at all like us, between what happened long ago and what is happening right now or could happen next. The urgency increases when the subject at hand, like war or disease, involves life and death, including the potential death

of all people on earth. We need to know, and we need to know something more than piles of unrelated observations.

So the reckless amateur rushes in. . . .[2]

[M]y suspicions first centered on the theme of gender. Wars, after all, have been fought, in almost all cases, by men, and sometimes for the stated purpose of "making" them men. But as I pressed back further in time and broadened my inquiry to include other forms of organized and socially sanctioned violence—such as rituals of blood sacrifice—a deeper and far more ancient explanatory theme emerged. This was not something I had ever thought about before . . . a new evolutionary perspective on war and related forms of violence. . . .

This is a theory about the feelings people invest in war and often express as their motivations for fighting—where these feelings might have originated and how they have played out in history.[3]

War . . . is too complex and collective an activity to be accounted for by a single warlike instinct lurking within the individual psyche. Instinct may, or may not, inspire a man to bayonet the first enemy he encounters in battle. But instinct does not mobilize supply lines, manufacture rifles, issue uniforms, or move an army of thousands from point A on the map to B.[4]

Proponents of a warlike instinct must also reckon with the fact that even when men have been assembled, willingly or unwillingly, for the purpose of war, fighting is not something that seems to come "naturally" to them. In fact, surprisingly, even in the thick of battle, few men can bring themselves to shoot directly at individual enemies. The difference between an ordinary man or boy and a reliable killer, as any drill sergeant could attest, is profound. A transformation is required: The man or boy leaves his former self behind and becomes something entirely different, perhaps even taking a new name. In small-scale, traditional societies, the change was usually accomplished through ritual drumming, dancing, fasting, and sexual abstinence—all of which serve to lift a man out of his mundane existence and into a new, warriorlike mode of being, denoted by special body paint, masks, and headdresses.[5]

Often the transformation is helped along with drugs or social pressure of various kinds. . . . Almost any drug or intoxicant has served. . . . So if there is a destructive instinct that impels men to war, it is a weak one, and often requires a great deal of help.

In seventeenth-century Europe, the transformation of man into soldier took on a new form, more concerted and disciplined, and far less pleasant, than wine. New recruits and even seasoned veterans were endlessly drilled, hour after hour, until each man began to feel himself part of a single, giant fighting machine.[6]

In the fanatical routines of boot camp, a man leaves behind his former identity and is reborn as a creature of the military—an automaton and also, ideally, a willing killer of other men.

This is not to suggest that killing is foreign to human nature or, more narrowly, to the male personality. Men (and women) have again and again proved themselves capable of killing impulsively and with gusto. But there is a huge difference between a war and an ordinary fight. War not only departs from the normal; it inverts all that is moral and right: In war one *should* kill, *should* steal, *should* burn cities and farms, should perhaps even rape matrons and little girls. Whether or not such activities are "natural" or at some level instinctual, most men undertake them only by entering what appears to be an "altered state"—induced by drugs or lengthy drilling, and denoted by face paint or khakis.[7]

In war men enter an alternative realm of human experience, as far removed from daily life as those things which we call "sacred."[8]

The emotions that overwhelmed Europe in 1914 had little to do with rage or hatred or greed. Rather, they were among the "noblest" feelings humans are fortunate enough to experience: feelings of generosity, community, and submergence in a great and worthy cause.[9]

The mass feelings inspired by war . . . are eerily similar to those normally aroused by religion. Arnold J. Toynbee . . . argued that war had in fact become a religion, moving in to fill the gap left as the traditional forms of worship lost their power over people. "Man," he wrote, requires "spiritual sustenance," and if man was now less inclined to find it in a church, he would find it in the secular state and express it as a militant nationalism in which "the glorification of war [is] a fundamental article of faith."[10]

By the twentieth century, war, and the readiness for war which is so much a part of nationalism, had become the force unifying states and offering individuals a sense of transcendent purposefulness.[11]

The Kingdom of God Is Green

> [T]here are at least two reasons to take seriously the religious dimension of war. First . . . it is the religiosity of war, above all, which makes it so impervious to moral rebuke. . . .
>
> The other reason . . . is for what it has to say about us as a species, about "human nature," if you will, and the clichéd "problem of evil." . . . [N]o other species exhibits behavior we recognize as "religious," and none can be said to bring exalted passions to their acts of intra-species violence.
>
> . . . What is it about our species that has made us see in war a kind of sacrament?[12]

> "Antecedent" may be a better word for what we are after here: Hunting is an antecedent of war, almost certainly predating it and providing it with many valuable techniques; [but] here we seek a similarly long-standing antecedent to the *sacralization* of war.
>
> . . . As Rene Girard emphasized in his classic *Violence and the Sacred*, violence was, well into the historical era, at the very core of what humans define as sacred, and the first question we will address is *why*.[13]

> [I]t is my contention that our peculiar and ambivalent relationship to violence is rooted in a primordial experience that we have managed, as a species, to almost entirely repress. And this is the experience, not of hunting, but of being preyed on by animals that were initially far more skillful hunters than ourselves.[14]

I am desisting with quotation after only two chapters from *Blood Rites* (there are fourteen chapters in all) because there simply is too much to quote. I urge every reader who has stuck with these present essays to carefully read Ms. Ehrenreich's book: it is not only about our common history, it is also very much about our patriotism, our nationalism, and our theology.

Barbara Ehrenreich traces our bloody rites back to very early human fears of predator beasts, and to human exhilaration at turning the tables. The flag—patriotism, nationalism—becomes the "secular" tribal emblem around which we gather to recharge our raw emotions against the Other. (This could well be why patriotic expressions contain such primal emotion.) And so this ancient human ritual builds in intensity and consequence as the contending Others (or Beasts) become overwhelming not merely in size but in technological lethality. Plus, it must be said, patriotism as a form of male bonding is culturally mimetic and self-perpetuating. Perhaps we might grasp Arnold Toynbee's designation "rival witch-doctor" (referring to ideological competition between capitalism and communism) as something more than a mere burst of wit, something in fact of primal potency.

Leopold Kohr wades into this discussion—I have a hard-to-find copy of *The Breakdown of Nations* at hand—by proclaiming that there's no solution, certainly no majestic supergovernment solution, to this problem. Human beings are human beings, he says, boys will be boys, and the only solution (though nobody will listen) is to deliberately restore smallness of scale in all political and economic units: not undoing thereby the human proclivity for conflict and war, but radically restricting its playing field and scope, therefore reducing its destructive magnitude.

One senses in Kohr a delightful companion and conversationalist, a wit, something of a bohemian (though also a serious academic), fully at home with good wine, fine food, and lively minds. With Barbara Ehrenreich one senses something different: where Leopold Kohr seems wittily willing to accept the human condition as he finds it (even as he criticizes it), Barbara Ehrenreich deeply desires some sort of cultural transformation (even if she's not sure such a transformation can be achieved).

I am predisposed to like Kohr; he's in the same broad "anarchist" camp as Martin Buber, E. F. Schumacher, Ivan Illich, Kirkpatrick Sale, and Wendell Berry—persuasive intellectuals who spent their lives attacking the Big (the "predator beast"?) in very direct and compelling books. I also like Ehrenreich for the troubled, open-ended humanity in her brilliant probings. But I'm not fully satisfied with either Ehrenreich or Kohr, and my dissatisfaction boils down to a matter of Spirit.

Certainly one can make the case (lots of such cases have been made) by which life and human history are without spiritual guidance or spiritual meaning, except for what humans subjectively impute. This is pretty much the operating principle of modern scholarship, fact-based and totally secular. Such a stance is, in many ways, collegial, self-correcting, and scientific. It has broken or simply refuses to look through the mythological 3-D glasses of hardened religious ideology: no six-day creation, no Noah's Ark, no parting of the Red Sea, no literal resurrection of Jesus. Supernaturalism is what Sunday school teachers and most preachers get to talk about from their harmless chairs and ego-inflated pulpits. Lots of mythological yackety-yak. Meanwhile the real world rolls on.

This clear-eyed scientific (and rather brutal) stance has let a lot of fresh air into the cathedral over the last several centuries, even if the cathedral tried to keep its windows closed and doors locked. Science has largely discredited a hugely conventional image of God, a god who never existed. But science has not, in my estimation, discredited Spirit: that Something too slippery, unmanageable, and elusive for conventional human consciousness to grasp or pin down. Discussing theology can be a lot like shoveling water with a pitchfork—an abundance of vigorous motion, a lot of splashing, and nothing much to show for the effort. If religion postulates an idolatrous theism as its central tent pole, scientific secularism postulates—even if it is implicit rather than explicit—an equally idolatrous atheism. The former has developed some deadly divine tribalisms—let's say "the Middle East," for an oblique but obvious example. The latter diddles with atomic weapons in the name of *Realpolitik*. Is there no way out of this box? Is it either fighting over God and ego or oil and ego?

Barbara Ehrenreich's deep analysis serves to make John Shelby Spong's assertion meaningful in a very scary and provocative way. I mean Spong's idea that theism has been the core organizing principle of human behavior since the dawn of human consciousness. Spong's theism and Ehrenreich's predator beast are suddenly, and clearly, two psychocultural artifacts of great human antiquity—two artifacts that, at the least, interpenetrate, and may even be one and the same. Theism may have at its core a leering predator beast. (What do we really have in a fierce God who, we are told, will throw all human refuse into a fiery pit of eternal torment? This is uncomfortably close to a Devil who stands on the rim of that same fiery pit, looking down with leering satisfaction.)

This linkage between predator beast and theism also begins to make explicable and meaningful another heavy-duty term that I've been trying, in these essays, to stuff back into its etymological box, a term whose relationship to its original meaning is similar to what urban sprawl is to "city." That word is *pagan*. Here, at last, may be an adequate "pagan" groundedness for "country district": an ancient human vulnerability to predator beasts in a wild landscape totally devoid of institutional securities, and then both the fear and the theistic projections so generated by deeply (and collectively) internalized dread. This ancient and deep fear of predators simply got transformed into *our* god who is the Lion who protects us; while *their* god, meanwhile, is the Tiger who seeks to devour us. Lions v. Tigers, our team versus their team, our kingdom versus their kingdom, our god versus their god—a sportive, tribal theism of predator beasts, a true theistic "paganism" of predaceous beasts, as openly and commonly

This Predator Beast Game

portrayed in the heraldry of contending sports teams: the Warriors, the Raiders, the Red Skins, the Bulls or the Bullets.[1]

I do not believe that Lao Tzu, Siddhartha, and Jesus—to name only the spiritual teachers with whom I have the greatest familiarity—were into this theistic predator beast game. I believe they saw into it, rejected it, and tried mightily to shift our psychodynamic spiritual concentration onto a different plane, a finer spiritual comprehension not saturated with *Alptraum*, nightmare, but composed of fiercely gentle trust in the ultimate benevolence of Spirit, irrespective of suffering or fear of suffering. (This may also be the internal psychodynamic of Gandhi's *satyagraha* or "self-suffering," for "self-suffering" obviously relies on the courage to be compassionately vulnerable in the face of political threat. If we also meditate on Gandhi's promotion of village-mindedness, while pondering the apparent truth that the precivilized agrarian village was largely devoid of weapons, then we might provisionally conclude, first, that the agrarian village had reached a level of cultural and psychological security that had largely exorcized the predator beast, but that, second, the feral males who, with their bloody weapons, imposed their predatory aristocracy on the villagers and created civilization, were, on a psychic level, restoring dread of the predator beast—only this time the feral aristocrats were themselves the successful impersonation of predator beasts.)

The first clue that this is so is that neither Jesus, Buddha, nor the Old Sage of Taoism was a promoter of kingdoms, empires, or power-driven civilization. These exemplars of compassionate vulnerability certainly did not play a new predator beast mind game of rival witch-doctors. They saw—I feel very confident about this in regard to Jesus—that empire was the monolithic form or expression of this ancient duality: *Our* empire is *our* Lion, the enforcing power of *our* god, who will crush *their* empire, their Tiger, and their god. Jesus just said no to this dreadful game. He refused to play. He chose the nonviolent, deeply cooperative *malkuta*.

The superb and sublime irony of overcoming this ancient, deep, and pervasive "paganism" is that such overcoming thrusts us into full dependence on Spirit; and this dependence demands and requires of us deeply committed servanthood on Earth and deeply reverential stewardship in Creation. One cannot stress this strongly enough: the only path out of this predator beast game is spiritual in nature, with trustful vulnerability and

---

1. *Webster's*, in its etymology of "beast," suggests the ancient Indo-European root *dwejes-to*, meaning "that which is feared." This seems to me to lend a great deal of credence to Ms. Ehrenreich's hypothesis.

cooperative truthfulness. We have to let go of using "our" predator beast/god as a protection against vulnerability and, instead, embrace vulnerability as the portal to Spirit. The spiritual logic, it seems to me, is inescapable. "Paganism" in the ancient sense of predator beast fear and worship, this ancient personified (or bestial) theism, must atrophy or we will perish through the cataclysmic technological projections of our fear, hatred, greed, and arrogance. Either a postcivilized spirituality deeply grounded in stewardship and servanthood or a theistic civilization-versus-terrorism apocalypse of predator beasts that will kill us all.

Leopold Kohr says there is no way out except the liberation and restoration of the small-scale. Barbara Ehrenreich would have us concentrate our war passions against war itself. I recommend repentance. Chris Hedges, in *War is a Force that Gives Us Meaning*, says that love "offers the only chance to escape from the contagion of war. Perhaps it is the only antidote."[15]

Perhaps all of us are right.

Notes

1. Ehrenreich, *Blood*, 2.
2. Ehrenreich, *Blood*, 2–3.
3. Ehrenreich, *Blood*, 3.
4. Ehrenreich, *Blood*, 9–10.
5. Ehrenreich, *Blood*, 10.
6. Ehrenreich, *Blood*, 11.
7. Ehrenreich, *Blood*, 12.
8. Ehrenreich, *Blood*, 12.
9. Ehrenreich, *Blood*, 14.
10. Ehrenreich, *Blood*, 15
11. Ehrenreich, *Blood*, 17.
12. Ehrenreich, *Blood*, 19–20.
13. Ehrenreich, *Blood*, 21.
14. Ehrenreich, *Blood*, 22.
15. Hedges, *War*, 168.

30

The Feminine Dimensions of God

FOR WELL OVER THIRTY years, I have known of and admired the Catholic Workers. And although I've come to expect an ethical doggedness from these folks—houses of hospitality in inner-city neighborhoods, voluntary poverty, refusal to pay taxes, repeated acts of civil disobedience particularly in regard to war and weapons, attempts at cooperative-communal farming—there's always been (it seems to me) a steady, theologically conservative obedience to religious hierarchy and doctrinal orthodoxy within Catholic Worker radicalism. But now I have reason to believe this reflexive obedience is being overtaken and dissolved by Gospel yeast.

Jeff Dietrich has a sketch of "the male psyche" in the March-April 2002 issue of *The Other Side* magazine. "Seminary in the Cellblock," a six-entry section of Dietrich's jail journal in 2001, eats quite a hole in "conservative" obedience. (Jeff and his friend Brian were in California's Kern County Jail—"with the rest of the incarcerated losers"—for protesting the U.S. missile-defense system at Vandenberg Air Force Base on May 20, 2001.)[1]

In his journal entry for June 3, 2001, Jeff centers explicitly on the "all-male brotherhood of war and hunting, violence and survival," the "deadly food of male power and prerogative." He ties together, in a few brief, swift paragraphs, the tough brotherhood bonding of ancient traditional cultures, male initiation, and the more contemporary male toughness "associated with ideologies of power and control: business, technology, law, religion, politics, and war."[2]

The Kingdom of God Is Green

Notice, please, that religion is *not* absent from this short list of power and control ideologies. Against this embedded and endlessly destructive male toughness, Jeff Dietrich quietly contrasts Luke's Gospel with its women and mothers, its hospitality and compassion. He tells us that the "Hebrew word for womb, *rahem*, is closely related to the word for compassion," and that women (as we all know) are "associated with serving, nurturing, feeding, healing, and compassion." These are—and here's the term to watch—"feminine dimensions of God."[3]

In the February 2002 issue of *Catholic Agitator*, a publication of Catholic Worker houses in Los Angeles, Jeff Dietrich (released from jail on November 19, 2001) interviews Phil Berrigan (released from prison on December 15, 2001) largely on the subject of "this current war." After saying that the Bush administration "had all sorts of warnings from the Russians and from German and Israeli intelligence about imminent attacks" and "did nothing," Phil Berrigan responded to Jeff Dietrich's question about the U.S. Bishops' statement on the war against terrorism. "Are you familiar with their position?" asks Jeff:

> Phil: Oh, yes. I'm afraid you're dealing with guys who are mostly managers—managers of property and income and what have you. I've heard them accused by reputable sources of being biblically illiterate and ill-prepared to make the moral judgments that are required when these crises come down. They cling to the just war theory and convince themselves . . . all seven American Cardinals committed themselves to what the U.S. administration is doing in this war.
>
> Agitator: What do you think would be an authentic Christian response to the current situation?
>
> Phil: The Christian response, of course, is no violent retaliation—none whatsoever. A Christian response would involve admitting that the U.S. has been the number one terrorist in the world and that we have ravaged third and fourth world countries all over the planet.[4]

I think it perfectly safe to say that current U.S. policy is a menu for the "deadly food of male power and prerogative," and that what we desperately need, hope and pray for, is a global outbreak of the "feminine dimensions of God."

Notes

1. Dietrich, "Seminary," 12, 13.
2. Dietrich, "Seminary," 12.
3. Dietrich, "Seminary," 12.
4. Dietrich, "Pray," 4.

31

Such Pureness of Heart

I DON'T KNOW HOW many kindly grandfathers there are on the American Left, but two of the more visible and impressive ones have been Noam Chomsky and Howard Zinn. A small book from each—Chomsky's *9-11* and Zinn's *Terrorism and War*—arrived with Gore Vidal's *Perpetual War for Perpetual Peace: How We Got to Be So Hated*, all courtesy of my local Peace Study friend, Bill Ritter. I've now read all three books, and I've written a short, friendly admonition to Gore Vidal. So I'd like to turn, briefly, to the grandpas.

What strikes me most strongly about these books, beyond their clear and compelling analytical perspectives, is the depth and groundedness of their humane morality. Chomsky's *9-11* returns again and again to the meaning and definition of "terrorism," underscoring and expanding on the comments in the *Extra!* piece I quoted from and commented on in an earlier essay. Chomsky has an awesome hypocrisy detector. He insists that terrorism is terrorism, no matter who perpetrates it, and to pretend otherwise is the perpetual cynical ruse of *Realpolitik*. Meanwhile, *Terrorism and War* repeatedly reveals Zinn's deeply compassionate nature, his commitment to universal equality, and his sorrow (he was a World War II bombardier in Europe) over the senseless death and stupid violence of war. His gentleness can literally be felt.

This is, of necessity, a purely subjective assessment. But I know of no other grandpas in this country, including every possible religious or political figure I can bring to mind, who radiate such moral groundedness and ethical integrity, such pureness of heart—at least since Martin Luther King. If this is true—it's true for me—what does it say when our most relevant grandpa prophets are secular Leftists and, probably, atheists by conventional understanding? It means (*viz.* Altizer) that the theistic God who never existed is nevertheless dead and that the Spirit who is infinitely alive goes where "it" is most congruent—to wondrous grandpas like Zinn and Chomsky—for how can Spirit feel at home except where "it" is ethically welcome? (God is apparently too busy sorting out the legalistic qualifications for Just Wars to have much time for Spirit, who may well be horrified by all wars.)

Both *9-11* and *Terrorism and War* are "interview" books. In Zinn's, Anthony Arnove asks Grandpa Zinn a "God" question. Here's Zinn's answer:

> Well, if, God were not neutral, then Bush would be gone . . . but I've never checked up really on God's neutrality.
>
> The invoking of God has actually been done by all American presidents. Given the so-called separation of church and state, it's really ironic that every president seems absolutely obsessed with calling upon God to bless us. Ronald Reagan wrote, "Perhaps no custom reveals our character as a nation so clearly as our celebration of Thanksgiving Day. Rooted deeply in our Judeo-Christian heritage, the practice of offering thanksgiving underscores our unshakable belief in God as the foundation of our nation and our firm reliance upon Him from Whom all blessings flow." But Bush is invoking God more than anybody else. And then when you join this to his talk about a "crusade" against whoever perpetrated the attack on September 11, then it puts the United States in a position of a crusading country with God on our side.
>
> It's interesting that God is brought into the picture when the government is doing great violence. Maybe it's when you are doing great violence that you desperately need some support. You're not going to get any moral support from any thinking person, but since God isn't thinking at the moment, maybe you can pull out God to support you. He certainly isn't around to contradict you.
>
> It's a pernicious thing to do. It takes advantage of the fact that a lot of people in this country believe in God, go to church, and think of God as a moral force. Then you use God to support the most immoral of acts, which is war.[1]

We should ponder long and hard on the "use" of God to "support the most immoral of acts." It is one of the linchpins of civilization's justifications and terrorism's double standard. (Do we recall Barbara Ehrenreich

The Kingdom of God Is Green

saying that Arnold Toynbee "argued that war had in fact become a religion, moving in to fill the gap left as the traditional forms of worship lost their power over people"?)

Chapter 6 of Chomsky's *9-11* is entitled "Civilizations East and West." Unlike *Terrorism and War*, *9-11* is based on a sprawling set of interviews, each independently conducted, mostly European. Except for a convenient lead-in phrase on the last page of Chapter 5 ("civilizing effects of the popular struggles of recent years"),[2] Chapter 6 is the only place we even get a glancing definition of "civilization." This is both unfortunate and, given Chomsky's semantic clarity on the meaning of "terrorism," unsettling. On occasion—as when Chomsky says "As for 'Western civilization,' perhaps we can heed the words attributed to Gandhi when asked what he thought about 'Western civilization': he said that he thought it might be a good idea"—it appears that Chomsky recognizes the sack of worms undulating within the living corpse called Civilization; but he apparently prefers not to open that sack, or can, or sarcophagus, because a person could spend the rest of his life dealing with the creepy semantic mess and dreadful linguistic stench.[3] For this investigation we perhaps need forensic pathologists with lifetime tenure in the fields of traumatic institutions and civilized diseases.

Again, and probably to belabor a point that, in truth, needs such belaboring, the Left—in general; I don't really mean Chomsky or Zinn here—the Left clings to "civilization" because it needs an overarching concept or construct by which to ground and anchor its social vision of sharing and stewardship. The paradox (if paradox it is) is that no civilization to date has existed without an aristocracy or a ruling class hostile to both sharing and stewardship. Civilization was never formulated as a project for securing servanthood or promoting stewardship. The aristocracy did to the peasantry and laboring class what civilization as a whole has done to agriculture and industry: constantly wring from them all possible "surplus" by means of devices ranging from brutal coercion to perfectly legal economic manipulations.

So the term the Left clings to for its supposed inherent moral goodness and self-evident humanitarianism—civilization—is a construct whose entire history has been turned against the Left's vision of a just and equitable society. This makes as much sense as a professed atheist clinging to God. In my estimation, to hope for a just *civilization* is to daydream of a just aristocracy, an oxymoron packed with five thousand years of unremitting contradiction. This is the cultural trap the Left is in, due to its eschewal of the spiritual.[1]

On the other hand, I do not really think that the Gospels' term—the "kingdom of God"—is exactly going to catch fire as a viable semantic alternative to "civilization." Such a term has, needless to say, an enormous load of deadly theistic and ecclesiastical baggage all its own. Semantically, it is welded to the monarchical image of God. So I am, in truth, devoid of an alternative term or concept to bridge the gap. This absence of an alternative term very definitely is a liability, just as the absence of a fuller myth by which to integrate the precivilized is a liability; but absence of a satisfactory semantic alternative is *not* an adequate reason to cling to a term and concept representing, from its inception, a unified system of exploitation and oppression. I mean civilization. Sometimes we just need to let go of the security blanket. We will find a new and adequate term as we discover, create, or are just open to a new and adequate cultural myth; and we will discover, create, or be open to that myth as we simultaneously examine, with minute attentiveness, our actual past and push past the elsewhere, bogeyman God and discover indwelling, life-giving Spirit.

When asked bluntly—"Do you think we are using the word 'civilization' properly? Would a really civilized world lead us to a global war like this?"—Chomsky sidesteps the wormy question, though his marking of "civilized world" tacitly signals his awareness of the problem:

> No civilized society would tolerate anything I have just mentioned, which is of course only a tiny sample even of U.S. history, and European history is even worse. And surely no "civilized world" would plunge the world into a major war instead of following the means prescribed by international law, following ample precedents.[4]

Like Gore Vidal, these grandpas don't take us back far enough in time. They don't give us a long enough or deep enough historical perspective. (I'm not saying that civilization is the absolute bedrock basis of our woes, for civilization congealed out of actual and existing human propensities; but civilization is the institutionalization of Class and War, the "permanent" congealing of "traumatic institutions," and, as such, is the

---

[1]. Just as the trinitarian tripod or triangle must be reconfigured as a mandala (Father, Son, and Holy Spirit to Mother, Father, Son, and Daughter), so there are four basic characterizations of human consciousness: body, emotion, intellect, and spirit. Spirit is not some fuzzy excrescence of intellect. It is not the brain in a state of hallucination. It is as distinct a characterization as body, emotion, or intellect, and it is a tragic crippling of all human dimensions to deny it its proper place and function.

actual and existing construct we are faced with.) This inadequate historical perspective on the part of Zinn and Chomsky is, with all the gratefulness and praise I will otherwise freely lavish on the grandpas, my only criticism. But it's not a small criticism. At some point, we really have to open the tomb—like Mary Magdalene, in John 20:11–15, who is described as going to Jesus' tomb, standing outside, weeping. The tomb is seen to be empty. Mary turns, blurry-eyed, to someone standing behind her, *supposing him to be the gardener,* to ask where they have taken her beloved. Jesus is the gardener. The supposition was absolutely correct. A new Adam and a new Eve require a new garden, a new groundedness in Creation, and therefore the resurrection of the peasantry. The "kingdom of God" is one of the names that garden—that "resurrection"—goes by.

When the vile tomb of civilization is opened, and the stench dissolved in sunlight and fresh air, we will all be freed to become resurrected gardeners. In the evening, on the porch, we will listen as the grandpas and grandmas tell incredible, implausible stories of how people used to hate and kill one another in the names of justice and love, in behalf of democracy and God. The children, curled in warm laps, will fall asleep in the aroma of flowers, pondering such bizarre fairy tales.

Notes

1. Zinn, *Terrorism*, 93.
2. Chomsky, *9–11*, 70.
3. Chomsky, *9–11*, 92.
4. Chomsky, *9–11*, 80.

32

God's Lifeboat

IN MY NECK OF the woods, "human nature" is conventionally considered to be, largely, a negative thing. People are rotten, greedy, and selfish, and any attempt to "improve" society will eventually crash on the immovable rocks of human nature—or as Kevin Philips instructed us in Chapter 7, "liberal eras often fail through utopias of social justice, brotherhood, and peace," while conservative abuses involve "worship" of markets, the "elevation of self-interest" over community, and belief in the "survival of the fittest." Such "progress" as there may be is limited to technological inventiveness, explicitly and demonstrable, and, more metaphysically ambiguous but still deeply believed in, the social and cultural movement called "development" that lifts people out of "mere humanity," out of primitivity and backwardness, and on into civilization. Progress is possible, in other words, for machines and savages; all other possibilities are improbable and unlikely. Machines can always be improved; savages secretly yearn to be elevated to the exquisite sensitivities of civility, although "human nature" is, apparently, a glass ceiling (or a stone floor) when it comes to amelioration of civilized "diseases" or "traumatic institutions."

One is as likely to hear this view expressed, in various magnitudes of sophistication, by a middle school teacher as a grocery store clerk, a policeman as a mechanic, a bus driver as a clergyman. The mythology of "progress," its application and its limitation, has become this pervasive. It's as reflexive as a Walk light at a busy intersection.

Apparently, when one becomes "civilized," that's the last stop in this life—though, to be fair, there may be gradations of refinement. The next step up (if one believes in this possibility) is heaven, but you have to die first to get there. And, also apparently, in order to get there—that is, to heaven—you have to believe the right things in the right way, be a member in good standing of the right religion, or maybe even of the correct sect within the right religion. Everyone else is going to have to do some fast talking at the gates of heaven, and their chances of jiving the gatekeeper aren't too bright. The alternative destination is not very attractive and is alarmingly permanent.

This slightly spiced-up compaction is, I believe, a pretty fair summary of our prevailing mythology. A crude symbol for the constraints of this mythology is the dog collar by which your dear pet will be shocked if it strays beyond a certain electronic radius from your house. When Arnold Toynbee, as we saw in a previous essay, wandered back and forth between the secular and the sacred in his attempt to describe the "voyage" of civilization, he was not being brilliantly insightful or daringly provocative. He was simply articulating the common mythology, modestly taking his place as a footnote to Augustine. Civilization provides God with the domesticating (or at least restraining) leash for his wild creature Man—an attempt by human beings to "rise above mere humanity—above primitive humanity, that is—toward some higher kind of spiritual life."

So here we are milling around inside the radius of our civilized dog collar, waiting somewhat restlessly for a higher kind of spiritual life to settle in on us. Human nature is, thanks to Original Sin, pretty bad. According to Augustine, God gives and sustains kingdoms, empires, and civilization as a kind of for-your-own-good dog collar. According to Toynbee, civilization lifts us up and out of mere humanity and leads us toward the communion of saints.

Class and War, as "diseases" of Civilization, are rooted in human nature. Technological progress has elevated the destructive inclinations of human nature—the lethal instruments by which we express and make manifest our selfish and greedy human nature—into apocalyptic dimensions. The industrial revolution permitted, encouraged, and facilitated the transfer of slavery from human beings to machines and enabled a far more intensive penetration of and extraction from nature's reserves. In that respect, the industrial revolution both deflected and democratized slavery. Energy-intensive machinery was the throat in the hourglass of civilization through which "backwardness" and "underdevelopment" were pumped into regimented urban-industrial consumerism. But, as Arnold Toynbee points out, the unequal distribution of goods has become an intolerable injustice, a "moral enormity."

Given the civilized etiology of War and Class "diseases," it is perfectly reasonable to be utterly pessimistic about the future.[1] In a book called *An Inquiry into the Human Prospect*, Robert Heilbroner, in his "Final Reflections," asks if there is "hope for Man?" His answer is grim: "without the payment of a fearful price . . . there is no hope":

> Rationalize as we will, stretch the figures as favorably as honesty will permit, we cannot reconcile the requirements for a lengthy continuation of the present rate of industrialization of the globe with the capacity of existing resources or the fragile biosphere to permit or to tolerate the effects of that industrialization. Nor is it easy to foresee a willing acquiescence of humankind, individually or through its existing social organizations, in the alterations of lifeways that foresight would dictate.[1]

Heilbroner said—in 1974!—that "The existing pace of industrial growth, with no allowance for increased industrialization to repair global poverty, holds out the risk of entering the danger zone of climate change in as little as three or four generations."[2] (How about a mere two generations?) He goes on to say, however, that the human prospect is not an "irrevocable death sentence," not

> . . . an inevitable doomsday toward which we are headed, although the risk of enormous catastrophes exists. The prospect is better viewed as a formidable array of challenges that must be overcome before human survival is assured, before we can move beyond doomsday. These challenges can be overcome—by the saving intervention of nature if not by the wisdom and foresight of man. The death sentence is therefore better viewed as a contingent life sentence—one that will permit the continuance of human society, but only on a basis very different from that of the present, and probably only after much suffering during the period of transition.[3]

We need, says Heilbroner, a "gradual abandonment of the lethal techniques, the uncongenial lifeways, and the dangerous mentality of industrial civilization itself."[4]

I. In a 1998 novel entitled *The Smithsonian Institution*, Gore Vidal has his wisest character insist, on page 252, that "The human race will kill itself. That goes without saying. The virus—us—will kill our host the earth, or at least make it uninhabitable for us."

The Kingdom of God Is Green

Although Heilbroner suggests that "post-industrial society" may turn toward the exploration of "inner states of experience," he confesses that many of the "possible attributes of a post-industrial society are deeply repugnant to my twentieth-century temper as well as incompatible with my most treasured privileges." He is not cheered by a return to tradition and ritual, to "communally organized and ordained roles."[5] He hopes that "future man can rediscover the self-renewing vitality of primitive culture without reverting to its levels of ignorance and cruel anxiety."[6]

But after hammering primitivity for its ignorance and cruel anxiety, Heilbroner then says this about our "prevailing attitudes":

> When men can generally acquiesce in, even relish, the destruction of their living contemporaries, when they can regard with indifference or irritation the fate of those who live in slums, rot in prison, or starve in lands that have meaning only insofar as they are vacation resorts, why should they be expected to take the painful actions needed to prevent the destruction of future generations whose faces they will never live to see?[7]

Shall we blame this acquiescence and relish, this indifference and irritation, on primitive ignorance and cruel anxiety? Or shall we be spiritually mature and confess our sins without diverting culpability onto a scapegoat, especially one as defenseless as "primitive"? Only we can abolish civilization's Class and War, and then only by taking responsibility for civilization's "diseases" as we quit hiding behind the metaphysical shield called Original Sin.

The cure for these "diseases"—and, so far as I can see, the only cure—lies in the realm of spiritual transformation, and (although I think and write from within the Christian heritage) I mean by this term to include all the world's religions that have at their center a Creator, a Spirit, or Tao of boundless compassion and earnest forgiveness. This cure will not only abolish Class and War, it will also remove the compulsion and terror by which Class and War have their power—that is, the cure will dissolve the predator beast who masquerades alternatively as God or Devil. Involuntary servitude will essentially end. The aristocracy of wealth and power will cease to exist. If "civilization" remains after this abolition, it will in no way resemble or tolerate the instruments of the Class-and-War dynamics of all civilizations to date, nor will it sanctify past civilizations on the bogus rationalization of fruit for the few or no fruit at all—as if civilization were the wellhead of some kind of spiritual life or of Creation's vast abundance!

God's Lifeboat

We will come back to these points after a brief excursion into "war crimes" and "prisoner of war" status.

II

Katharine Temple has been one of several Associate Editors of *The Catholic Worker*, the New York City mothership newspaper of the Catholic Worker movement. In the March-April issue for 2002, Ms. Temple has a full page devoted to "Lessons from Nuremberg." Her comments open and close the page; in-between is her translation, from the French, of radical Protestant sociologist Jacques Ellul, as taken from an article or essay in *Verbum Caro*, of August of 1947.

The intent here, says Katharine Temple, is to raise the "question of justice for alleged perpetrators of violence (for instance, Augusto Pinochet, Slobodan Milosevic, Osama bin Laden, and some would add Henry Kissinger and the present Bush administration) through international law." But, it is "not easy to find substantial analyses of international law that are neither incomprehensible due to legalese nor suspect due to propaganda."[8] And so Ms. Temple brings forward this "closely argued piece" by Jacques Ellul. Ellul's critique, Ms. Temple asserts, is

> ... one of the sharpest contemporary, critical reviews of Nuremberg, not designed to make him popular. Even now, I do not expect unanimous agreement with his argument, either theoretically or practically. I present it for two reasons—as a historical study so we can assess our own situation more realistically, and also because he raises the most important questions about international law, questions that have yet to be answered.[9]

The core of Ellul's remarks (although not all of the *Verbum Caro* piece is reproduced in Ms. Temple's translation) is that the Nuremberg trials of Nazi war criminals presented a real dilemma to the allied judges. On the one hand, they wanted to directly address the "three main charges" of War Crimes, Crimes against Peace, and Crimes against Humanity. On the other hand, they wanted—quoting Ellul here—to "safeguard the sovereignty of nations." If we look carefully, says Ellul, we

> ... see we are in the presence of the defeated being executed by the winners. It has nothing to do with civilization condemning crime or war, but only the stronger doing what they want to the weaker. That is why we cannot speak of a division into good and evil, but only between conquerors and conquered. Was it

possible that this relationship would become just? Or, that this sentence would be an element in the establishment of justice?

This is the real problem with Nuremberg.

We do not have to be scandalized at the assertion that, at Nuremberg, there was only a relationship of force and the expression of vengeance, for this is indeed the state of affairs that provides the starting point for all law. We should be scandalized at the spectacle that says, "This is justice" where there is violence, or "This is right" where there is vengeance. This situation of violence and vengeance has to be surpassed; we should consider ourselves as being at the source of law. We have to ask what conditions are necessary for law to be worked out; if those necessary conditions are currently in place; if this demand for law does not challenge more than war and peace, but goes further to pose a decisive question for human beings, pushing us into an impasse?

We come back to what has been called a crime against humanity. The abomination of concentration camps, torture, executing hostages, deportations, bombing civilian populations, total war, looting of conquered peoples—all this is well known, but what exactly does it represent? It is a bit simplistic to see it as the result of dictatorship (though that would also be true), a particular or national sadism, etc. What makes it so immense is that it is not a single episode conditioned by politics or war. All this murderous activity is based on a conception of the world that comes directly from the givens of our civilization.[10]

Pull Ellul's two key thoughts together—violence and vengeance have to be surpassed; violence and vengeance arise from a conception of the world rooted in the givens of our civilization—and you have the core recognition and central plea of this present book.

The Nazis, Ellul insists, only "pushed to logical conclusions ideas contained in basic principles that are universally accepted." And then we get to the heart of the matter:

> In order for the Nuremberg verdict to be valid, it should have applied not only to the visible consequences, to the obvious scandal, but also to the causes. It should have called into question not just nazi concentration camps, but the concentration camp itself, including those of Russia, Spain or France. It should have called into question not only antisemitism but racism, including that of England and the USA—and so on, up to the values of a civilization that manufacturers these widespread facts. It would

appear obvious that it is a spiritual attitude that has made these crimes against humanity possible.[11]

And so we return to the spiritual question at the center of all these "widespread facts." Are we dealing with committed honesty or glib evasion? Straightforward confession or spin-doctor hypocrisy? What we fail to grasp—though the state of the world should clearly inform us—is that evasion and hypocrisy have consequences and are historically cumulative. They add up. They accumulate. These "widespread facts" blare out at us from every news broadcast. The world is now fully loaded—overloaded—with such manufactured facts.

III

Steve Rendall's article, "Take No Prisoners," appeared in the April 2002 issue of *Extra!*, which also contains Noam Chomsky's "Journalist from Mars," quoted from and commented on in an earlier essay. We will skip ahead more than half a century from Nuremberg to what Rendall calls "the Bush administration's defiant refusal" to grant prisoner-of-war status to "Afghan battlefield captives."[12] Here are prisoners, taken to the U.S. military base (Camp X-Ray) at Guantanamo Bay, Cuba, who are called "unlawful combatants" by the Bush administration.

Rendall tells us that the term "unlawful combatants" is "not a category from international law," but

> . . . a phrase used in a controversial 1942 Supreme Court ruling that denied regular trials to German saboteurs. (The saboteurs were given secret military trials in part because FBI director J. Edgar Hoover wanted to hide FBI bungling in the case—Charlotte Observer, 12/9/01.) The effect of treating the Afghan detainees as neither POWs nor criminals is to remove the captives from any established justice system, under either international or U.S. constitutional law—particularly when they are held at Guantanamo, where a federal appeals court has ruled that non-U.S. citizens have no "cognizable statutory or constitutional rights" (Broward Daily Business Review, 1/31/01).
>
> Critics say the administration has no right to unilaterally deprive detainees of POW status—that many, if not all, likely qualify under Article 4 of the Third Geneva Convention, which defines POWs as "members of the armed forces of a Party to the conflict, as well as members of militias or volunteer corps forming part of such armed forces." If a prisoner's status is in

question, he or she is entitled to a court hearing, according to Article 5 of the Third Geneva Convention: "Should any doubt arise . . . such persons shall enjoy the protection of the present Convention until such time as their status has been determined by a competent tribunal."[13]

The bulk of Rendall's article is a documentation of the extent to which mainstream U.S. media glossed over the Bush administration's "arbitrary designation" of the captives as "unlawful combatants." What we have here, it seems to me, is a glaring example of what Jacques Ellul called, with Nuremberg, the desire to "safeguard the sovereignty of nations"—namely, in this instance, the freedom of SuperPower U.S.A. to act unilaterally at will, and to take or leave international law when it fits or doesn't fit strategic purpose. Even more fundamentally apt and to the point is Noam Chomsky's critique of "moral truisms," that is, that hypocrites are "those who apply to others the standards that they refuse to accept for themselves," and who also impose on others the designations they refuse to accept for themselves. (Or is it true, as Robert Heilbroner says, that "There seems no hope for rapid changes in human character traits that would have to be modified to bring about a peaceful, organized reorientation of life styles"?)[14]

IV

The industrial destruction of the peasantry by enclosure and eviction from the commons, the forcing of peasant society through the factory-made machinery of the civilized hourglass, has not brought about utopian "civility," at least not as the term is commonly understood. Not only have we not reached a man-made paradise, we are in process of creating a toxicological disaster on Earth and of altering Earth's very climate even as our "weapons of mass destruction" become regular features on the nightly news, along with grisly demonstrations of desperate "terrorism" and unprecedented Antarctic ice floes. At the eye of these converging hurricanes stands cool human arrogance, dressed in the full, elegant hypocrisy of civilized values—and, at his side, the ministers and priests of evasion, draped in the solemn robes of sanctification. The "voyage" of civilization concludes in a global whirlpool of horrific violence and ecological destruction—a viciously ironic twist to the classical conceit of fruits for the few or no fruits at all.

If we are going to keep our metaphors temporarily afloat, the arrogance of aircraft carriers will have to be abandoned and sunk, and our lifeboat powered by wind, sun, water, oars and paddle. It's not nearly enough to assert the negative aspects of "human nature" and then act in exactly those ways that both demonstrate and prove the assertion: we have to repent of our assertions, our demonstrations, and our proofs. We have to let go of our enabling hopelessness and turn off the glib, well-funded prevarications.

In Matthew 8, Mark 4, and again in Luke 8, Jesus, asleep in a boat, is roused by his friends in the midst of a squall that threatens to capsize the boat and drown them all. Jesus, in these accounts, wakes from his sleep and rebukes the wind and rough water, which then subside and become calm.

"Believe" this story at whatever level of metaphor speaks to you. But at the level that speaks to me, I find that only the spiritually cleansed are finally worth listening to and following, that only they see clearly into the potentials of life, and those who thunder their asinine authority from the guarded podiums of Class and War remain wrapped in the secular religion of hypocrisy and warped by the religious insurance of evasion.

I find that I continue to believe in Spirit even as I have grave doubts about theism. I pray, though mostly with groanings of spirit. Recently I read a fairly long article by Robert Kaplan on Samuel Huntington, the "clash of civilizations" scholar. Huntington was, in passing, referred to as an Episcopalian, yet his political science doctrine is explicitly a doctrine of organizational power, of *Realpolitik*. (Perhaps, in the ecumenical spirit of Henry VIII, Machiavelli is the patron saint of Episcopalians.) So, is Jesus just a revolutionary Mad Max, an anarchist who, by a combination of great recklessness, courage, and peculiar historic accident, somehow got transmogrified into his virtual opposite? So that now a political scientist, who is an Episcopalian, can simultaneously be a "Christian" and an explicit advocate for the power politics of empire?

I am somehow compelled to throw my hat in with the Catholic Workers, the Quakers, the "peace churches" in general, the Buddhists—believing that there is a level, a dimension, and a dynamic of Spirit—though some may call it God, Great Spirit, Tao, or Allah. In the end, it's not really that important what the name is, so long as we are not really worshipping the projected psychodynamics of male hostility and righteous presumption. But I cannot live at or with the level of smug entitlement that simply saturates civility. I cannot escape the conviction that if we truly

The Kingdom of God Is Green

are creatures of cosmic accident, only complex chemical material bristling with intelligence but lacking in compassionate soul, we are, as a species, doomed: doomed precisely because our intrusive, selfish intelligence has created—and will continue to create—instrumentalities that, as a species, we are (or will be) ultimately unable to control. (The only option here is a now uselessly retrospective one, exemplified by Edward Abbey and the EarthFirsters, of wishing we had stayed at the hunter-gatherer level of cultural complexity.)

Transformation or disaster—so far as I can see, those are our only real options. Neither materialism nor idealism has adequate spiritual traction by which to open transformative doors. Therefore I conclude: if there's no Spirit, we are doomed; if there is a possibility of transformation, it's because there is Spirit. (Whether that's "theism," I honestly don't know and mostly don't care.)

But I do know that I increasingly feel a kinship toward those people who also have this sort of spiritual intentionality, all the way back to universal teachers like Lao Tzu, Buddha, and Jesus. I think they all, whatever the particularity of their conceptual language, were engaged in discovering, clarifying, and promoting the "kingdom of God." Therefore religion is the Great Nut to be cracked: not for the superficial worship of the shell, but for the communal sharing of the ethical substance inside.

The "cure" for the "diseases" of civilization is personal and political conversion to servanthood and stewardship. The abolition of Class and War opens the way to the "kingdom," the "realm," the "village," or the lifeboat of God.

Notes

1. Heilbroner, *Inquiry*, 136.
2. Heilbroner, *Inquiry*, 128.
3. Heilbroner, *Inquiry*, 138.
4. Heilbroner, *Inquiry*, 138.
5. Heilbroner, *Inquiry*, 140.
6. Heilbroner, *Inquiry*, 141.
7. Heilbroner, *Inquiry*, 143.
8. Temple, "Lessons," 4.
9. Temple, "Lessons," 4.
10. Temple, "Lessons," 4.
11. Temple, "Lessons," 4.
12. Randall, "Take," 17.
13. Randall, "Take," 17.
14. Heilbroner, *Inquiry*, 131.

33

The Global Cloning of Civilized Desire

IN REAL HISTORICAL TERMS, it's possible to identify four steps or stages, each packaged in its own pseudosacred rationalizations, which have brought the world to its current state of accelerating and deepening crises. These four are: the War and Class origins of civilization; the Constantinian Arrangement whereby Caesar and Jesus supposedly joined hands in the common cause of civilizing the pagan world; the twin, virtually simultaneous bursts of democracy and industry whereby modern civilization theoretically became democratic and supposedly abandoned slavery; and the current globalization of aggressive civilization that both evokes terrorist resistance and provokes ecological breakdown.

Civilization, in its rise from local feral bandit to global managerial CEO, has not only justified itself at every turn, it has also thrust itself and its justification into the very texture of human culture and into the ether of political consciousness. It is the single most successful virus of propaganda and propagation the world has ever seen. This core idea is remarkably easy to articulate, but nearly impossible to assimilate, due, largely, to the sheer immensity of its ramifications. Nevertheless, let's try to connect some dots and see what sort of picture we get.

First, behind Toynbee's apologetics for the armed perimeter of aristocratic luxury (fruits for the few or no fruits at all) lies the formation of civilization proper, with its "diseases" of Class and War. (And from John Crossan's witty little remark about how aristocrats don't "consider themselves 'above' peasants but beyond them somewhere," we can extract the obvious historical truth that aristocracy had no such "trustee" mandate

The Kingdom of God Is Green

as Toynbee would like to believe. Aristocrats were not acting with heroic intent as a vanguard for democracy.) And behind the formation of civilization lie a few thousand years in which gatherers, in particular, transformed botanical gathering into horticulture, thus enabling the settled and stable village to emerge, with domesticated plants and, then, animals. And behind this village culture lie, perhaps, hundreds of thousands of years of largely nomadic life of small bands of hunters and gatherers—the great historic reservoir of all human culture, huge in its adaptability, depth, and complexity. If civilization is our slave-owning parent, the precivilized agrarian village is our legitimate grandparent, while the bands and tribes of gatherers and hunters constitute our universal great-grandparent—older and far more venerable than Methuselah.

One could say that civilization, up until the industrial revolution, kept the peasantry at its precivilized agrarian village level; but, more precisely, the culture of the peasantry was essentially frozen at a level less flexible and less creative than that attained by the agrarian village prior to its impoundment by civility. That is, the aristocracy, historically, by violence and the threat of violence, systematically expropriated the production of the peasantry so that it might, in Toynbee's words, serve as a "kind of trustee for all future generations of the whole human race." But this alleged trustee status, imposed on peasants by the "dominant minority," prevented the agrarian village from whatever form or forms of cultural evolution it would otherwise have explored. Civilization has stifled the cultural evolution of the agrarian village for literally thousands of years.

Obviously, I consider the core of Toynbee's assertion to be a lie of historic proportions. Not that Toynbee was a liar—I don't suggest that for a moment—only that his conventional doctrine is fundamentally false, an articulation of pseudosacred rationalization all too common to civilized intellectuals. From the vantage point of the Christian Gospel, civilization is the system of oppression against which the kingdom of God stands opposed—powerlessness against power, honest subsistence against stolen opulence, the pruning hook versus the spear. Furthermore, what William Stringfellow calls the "Constantinian Arrangement" institutionalized the delusion that Church and Civilization could enter into a binding contract not only beneficial to both parties, but uncorrupting to both parties. It is this delusion Toynbee promotes with such elegance: civilization as an effort toward some higher kind of spiritual life, a Communion of Saints on Earth.

The Global Cloning of Civilized Desire

If the historic justification for aristocracy is the first lie, the Constantinian Arrangement is the second lie. The third lie and fourth lie have unfolded in a closer time frame. The third lie is that civilization can be, or to some extent already has become, "democratic," that it has abandoned and renounced slavery, that the institutions of Class and War either are now (somehow mysteriously) democratic or reflective of the democratic spirit. In other words, civilization, with its Class and War, is "democratic" because we've agreed to say that it is, because we have elections and factories; and, if we agree, it must be true.

The fourth lie emerges from all previous lies—from the historic justification of fruits for the few or no fruits at all, from the supposed marriage of Gospel and Empire, from the "democratization" of civilization—and is the most deadly, the most voluntarily banal, of all the lies. In the evaporation, meltdown, corrosion, and corruption of noncivilized folk cultures, civilization in its globalized form infiltrates the very consciousness of women, men, and children the world over with its astonishingly infective images of "development," of stuff, of goods, of consumer items, of the Utopian Happy Life that supposedly comes with "development." The global cloning of civilized desire into every nook and cranny of "backwardness" is the fourth lie, an immaculate deception in a heavenly credit card of plastic/electronic instantaneousness, while the doctrine of Original Sin makes us all equally responsible for any systemic failure.

If we dare, for a moment or over the course of a lifetime, to take Ivan Illich seriously and accept his view that this is *not* a "post-Christian world," that what he calls the "conglomerate of a series of perversions"[1] is, in fact, what we might more accurately call the poisonous flowering of Christian civilization, then—if we can hold to and be held by the simple moral truisms of the Gospel—then what?

Then what? is the question we all face. This is the pervasive, and perhaps even desperate, spiritual angst we need to cultivate in our souls. The test and measure of this angst is whether it results in our renunciation, in whatever ways, of the conglomerate of perversions, or whether, with a shrug, we turn toward the mercury vapor lights of Happy Life and willfully enter spiritual oblivion. Original Sin culminates in universal catastrophe. It is that simple. It is that hard.

Notes

1. Cayley, *Corruption*, 47.

34

The Gardener from Amenia

LEWIS MUMFORD HAS A little note on the title page of "Utopia, the City, and the Machine," as it appears in the paperback edition of *Interpretations and Forecasts: 1922–1972*. That little note informs the reader to look in his 1970 book, *The Pentagon of Power*, especially in chapter 8, "Progress as Science Fiction," for "further confirmation" of the "significant problems overlooked by the classic utopian writers."[1]

Mumford is dear to me. He should be dear to all of us. His insight that civilization is inherently utopian, that it always and everywhere has sought not merely to transcend nature but to *replace* nature with a superior man-made substitute is a profound and revolutionary insight. No historian I know has been more genuinely prophetic than Lewis Mumford: not "prophetic" via some unsubstantiated clairvoyance, but by a deep and pervasive understanding of historical dynamics. Mumford knew history so well he could read its underlying "archetypes." He saw our disaster coming. And he pointed a way out.

He also was a dedicated backyard gardener from Amenia, New York. He loved to have his hands in dirt. He loved to see plants grow. He loved nature, and he never lost sight (as way too many intellectuals do) of how totally dependent we are on the natural world and on the cultural achievements created and developed prior to the civilized imposition. Mumford loved history, loved the wonderful achievements of the human race; but he always knew, and he always kept in mind, that human achievement, in the end, must be compatible with both nature and human nature. Civilization, said Mumford, has violated both.

The Gardener from Amenia

In *The Pentagon of Power* (fourteen chapters with an Epilogue), Mumford tucked in two clusters of black and white photographs. Mostly they are stunning illustrations of tendencies ("Power," "Speed," "Remote Control," etc.) that are accelerating and intensifying the destructiveness of globalized civilization. In the aftermath of September 11, two photographs especially jump out.

The first is an actual aerial photograph of the Pentagon and the second is either a photo, or a photo of a mockup, of the World Trade Center. About the latter, Mumford says it is characteristic of the "purposeless giantism and technological exhibitionism that are now eviscerating the living issue of every great city."[2] The former, meanwhile, is a "symbol of the absurdity of totalitarian absolutism," an "imperviousness to information coming from outside sources and expressing human desires and purposes that have no status in the power complex."[3] And then Mumford says something that, forty years after the publication of his book, should cause us to sit up and take notice: this imperviousness "itself helps explain, perhaps, the increasingly desperate human reactions that the system is now provoking throughout the world. Never before has such a vast number of human beings, virtually the entire population of the planet, lived at the mercy of such a minuscule minority, whose specialized knowledge seems only to increase the magnitude of their incompetence in the very areas of their professional specialization."[4] Note that the embodiment of this imperviousness—the Pentagon—is supposedly the agency of a *democratic* America.

In the recommended eighth chapter of *The Pentagon of Power*, Mumford underscores the point: "All civilizations had carried with them for some five thousand years, I emphasize again, the traumatic institutions that had accompanied the rise of earlier power systems: human sacrifice, war, slavery, forced labor, arbitrary inequalities in wealth and privilege."[5] But we, of course, because we are wholesomely democratic and profoundly Christian, have somehow magically shed the traumatic institutions. *Our* civilization is a shining light on a hill, a beacon of pure motives devoid of blemish or disease.

There is no denying that civilization has been credited with a range of developments we would be loathe to do without. Whatever the bloody, oppressive, destructive process whereby we have, however tentatively, arrived at a single world, or at least the possibility of such a world, we now do really wish to sustain globalization in its finest features. Whether or not the greedy, murderous, deranged men, made wild by visions of conquest, fame, glory, and booty, *intended* to create a planet of humane cultural reciprocity is important but not crucial. What *is* crucial is that we have reached the point where greed, murder, and doctrinal fanaticism—religious or secular—have to be disarmed and discontinued, without equivocation or delay. Manifest Destiny is over.

There is no pill to cure human nature, but deference to egotistical madness must end. Disarm terrorism and terrorists? By all means. But we have to be spiritually courageous and begin a process—a determined process—to disarm, as well, state or civilized terrorism: what Mumford calls "traumatic institutions." Those who claim such disarmament is impossible or "unrealistic" are, ultimately, siding with the murderous status quo. Meanwhile, this current crisis—September 11 and its aftermath—tells us that to "go after terrorists," without a thorough soul-searching on the part of all who claim to be "civilized," is only to spin violence globally out of control.[1]

The real questions are astonishingly basic. Now that we have achieved something resembling global consciousness, recognizing in principle that all people are brothers and sisters, are we willing to share equitably with them? not exploit them? work to achieve something of a global standard in human rights, healthcare, education, economic security, and ecological coherence? Now that we have glimpsed the garish horizon of man-made global catastrophe, recognizing in principle and in fact that the utopian standard of living is ecologically cataclysmic and culturally pathological, are we willing to voluntarily reduce our consumption (proportionate to our advantages) and work, hard, to learn what it means to live simply in ecological community?

If the answer is *no*, the future—such future as there may be—will be increasingly bleak and destructive, merciless and cruel.

But if the answer is *yes!*, then our children and grandchildren will be privileged to live in one of the friendliest, warmest, most culturally creative periods in human history.

To be a true conserve-ative, to delight in the health of plants and animals, to wander in wonder in Creation and to puzzle over the apparent truth that all consciousness—all life—is also *self*-conscious, is to discover

---

[1]. Gil Bailie, in his *Violence Unveiled*, pages 63 and 64, while stressing that there is "moral difference between the violence of those trying to restore order and restrain criminal behavior and the violence of criminals, sociopaths, and terrorists," also says this "moral distinction is no longer categorical"; the "necessary asymmetry between 'good' and 'bad' violence is breaking down." That this "asymmetry" is breaking down at precisely the point of civilized globalization should be instructive.

The Gardener from Amenia

the sublime mysteriousness of multitudinous beings alive in Being. To live in this state of wonder, to live simply but beautifully in ecological community, to live with compassion, reverence, and humility, is to draw near to all true spiritual teachers and teachings—to live, perhaps, in what Jesus apparently called the kingdom of God. Or the *malkuta* of the Goddess.

The precondition, the doorway, for all this seems to be the relinquishing of utopian male presumption, contrite withdrawing from a male imperialism that seems to rest on a perverse blend of gender estrangement and its magical transmutation into aggressive superiority and resentful supremacy. This necessitates a huge step into trust, into the renunciation of violence, a deliberate willingness to endure vulnerability and an honest seeking for sexual and racial reconciliation. Those who truly know the Creator, the Tao, the Great Spirit, can go freely and easily through this doorway.

The door is small, of human scale, but there is room enough for everyone. May we all bow to enter.

Notes

1. Mumford, *Interpretations*, 241.
2. Mumford, *Pentagon*, 340-d.
3. Mumford, *Pentagon*, 180-g.
4. Mumford, *Pentagon*, 180-g.
5. Mumford, *Pentagon*, 199.

35

Pushing One Hundred

IN 2009, THE PROGRESSIVE was one hundred years old. In mid-April, 2004, the May issue of this beloved, cantankerous magazine—always a little ahead of the curve, perhaps by a century or so—made its appearance in my mailbox, with a colorful 95 on the cover. Inside was the usual bright and perceptive array of articles and commentary by some of the most tender and truly wise people living in America. Molly Ivins (now deceased), with her perfect Annie Oakley Texas quick-draw markswomanship. Bill Moyers, with his long-term earnest repentance for being LBJ's altar boy. Howard Zinn, with *his* long-term repentance for what he helped do to certain European cities from the bomb sights of an Army Air Force bomber. Nat Hentoff, with his endless, tenacious defense of the First Amendment. Noam Chomsky, whose brilliant acute rationality probes the stubborn mud of American willful stupidity. And, the current editor of *The Progressive*, hanging in there against all odds of being made irrelevant by the End of History (or other such premature pronouncements), Matthew Rothschild.

This essay will be, perhaps, the final contribution to *The Kingdom of God Is Green*. A portion of the manuscript has been sent to yet another publishing house (this one in St. Paul, home to Lake Woebegone and other cultural landmarks, like the SEAL-yell governorship of Jesse Ventura), and, as I remarked in my cover letter to the publishing concern, my analysis seems to fall in the cracks between the secular and the religious: too spiritual for the fact mongers, too Earth-based for the conventionally pious.

Now *The Progressive* is not given to religious tracts or spiritual exhortation. But here is Bill Moyers from his piece, "Our Story." These are the concluding paragraphs of Moyers' opening anecdote about a story he wrote, as a young East Texas reporter, about some well-to-do women who decided not to pay Social Security withholding for their domestic workers:

> Those women in Marshall, Texas . . . were not bad people. They were regulars at church, their children were my friends, many of them were active in community affairs, their husbands were pillars of the business and professional class in town. They were respectable and upstanding citizens all. So it took me a while to figure out what had brought on that spasm of reactionary rebellion. It came to me one day, much later. They simply couldn't see beyond their own prerogatives. Fiercely loyal to their families, to their clubs, charities, and congregations—fiercely loyal, in other words, to their own kind—they narrowly defined membership in democracy to include only people like them. The women who washed and ironed their laundry, wiped their children's bottoms, made their husband's beds, and cooked their family meals—these women, too, would grow old and frail, sick and decrepit, lose their husbands and face the ravages of time alone, with nothing to show from their years of labor but the crease in their brow and the knots on their knuckles. Even on the distaff side of laissez faire, security was personal, not social, and what injustice existed this side of heaven would no doubt be redeemed beyond the Pearly Gates.
>
> In one way or another, this is the oldest story in America: the struggle to determine whether "we, the people" is a spiritual idea embedded in a political reality—one nation, indivisible— or merely a charade masquerading as piety and manipulated by the powerful and privileged to sustain their own way of life at the expense of others.[1]

Well, there's the question: Is democracy a "spiritual idea embedded in a political reality" or merely a "charade masquerading as piety"?

I believe these contending options—embedded spiritual idea versus pious charade—constitute our core political dilemma. In the Christian religion, it boils down to the Sermon on the Mount versus John 3:16: the hard, rich, rewarding promise of an *ethical* spirituality that leavens, eventually, the entire political loaf, versus the obsessive mental concentration of a transcendent *believing* spirituality that promises to levitate the believer, eventually, into an eternal afterlife of bliss. The former is inherently Earth-based, Creation-centered, trusting truly that Spirit *is* love and that the "kingdom of God" truly *is* about the slow but steady yeasting of this love (manifested as servanthood and stewardship) throughout human consciousness and fully into human institutions. The latter is Heaven-oriented, salvation-yearning, fearful that unless the Right Belief is accurately arrived at and firmly adhered to, God the Judge will toss the failed believer out with the worthless chaff and the ill-tempered goats, into the fiery pit of eternal torment.

This latter view, even when secularized as the deserved fate of Backwardness, is what has us in its grip. It is the very stuff, the energizing current that has enabled or empowered civilization, with its murderous "traumatic institutions," from its founding. Its presumptions have become so exceedingly normative, so self-evidently obvious, so deeply a part of everyday consciousness and expectation, that the counter view, the "spiritual idea embedded in political reality" view, is seen as bizarre—as bizarre and as irrelevant as organic gardening is to genetic engineering, as building your own home is to a mortgage banker, as home birth is to your cyberspace HMO—at best a harmless, private atavism, but, when advocated in public, an aggravation to be rectified by law.

Civilization has sought to overpower nature and rise above Creation. The technologies associated with the industrial revolution (especially those developed and deployed since World War II) have permitted an astonishing new magnitude of overpowerment and rising above. The transcendent ideal for which civilization hankered, and for which state-protected religion claims to have the inside dope, has now been powerfully technologized and is, largely, available for private purchase and institutional procurement. You too can reside in Utopia with your plastic credit card, with your afterlife enrollment in Eternity.

This triumph is a colossus in a self-made, collective whirlpool of disaster, leaving no part of life on Earth untouched. It is going to get worse—much worse—before it gets better, and better will only come when we abandon our charades masquerading as piety, that is, when we voluntarily let go of our arrogant prerogatives and submit to live lives of humble stewardship and compassionate servanthood on Earth. And that is to say: the ecological crisis is a cultural crisis, which is an economic and political crisis, which is a spiritual crisis. Noam Chomsky, in his *Progressive* interview, says "Somehow the fact of enormous privilege and freedom carries with it a sense of impotence, which is a strange, but striking, phenomenon."[2] This privilege is our prerogative, our piety charade, and also our impotence.[1]

---

1. Lawrence Goodwyn, on page 318 of *The Populist Moment*, talks of "mass political

The combination of religious obsession with Eternity and civilized obsession with utopian perfection has produced a technological consciousness and lifestyle that finds Earthlife largely irrelevant and boring. (Boredom is an aristocratic disorder, and when common people are universally afflicted, it means that, at last, the spiritual sickness of boredom has become contagiously democratic.)

There may be life after death. I do not argue against it. (We all shall find out soon enough.) But the kingdom of God is among us and in our midst. This is the vineyard in which we are to make good juice and fine wine. Those who are most stridently for salvation are, I'm sorry to say, the most deluded, the most obsessed with the charade, the most willing to destroy this portion of Creation for the sake of their ideological preoccupation—their obsessive pursuit of immortality and their enchanted enthrallment with perfection. The proposal that we might redeem injustice this side of the Pearly Gates is, for them, laughable if not an obscene heresy. Utopian perfection has only dismissive contempt for eutopian wholeness.

The kingdom of God looks bleak, boring, and tawdry to those whose imaginations are addicted to salvation, Utopia, and civilized playthings. The reign of servanthood and stewardship lacks flash and fantasy. The kingdom of God is wonderfully ordinary. Like the Tao, like water, it seeks the lowly places and is content.

The kingdom of God is damp and Green.

Notes

1. Moyers, "Our," 29–30.
2. Chomsky, "Progressive," 39.

---

alienation" in the twentieth century, how that century "may well become known as the century of sophisticated deference." (I quote Goodwyn more extensively in Chapter 10, "The Underlying Religion of Civilization," in *Polemics and Provocations*.) But we need to recognize that contemporary deference, sophisticated or otherwise, is the product of bewildered alienation, and alienation is the inevitable result of the loss of self-sustaining folk culture combined with a nearly total immersion in utopian civility.

36

One Last Thing

I WISH TO SAY one last thing about the two kingdoms. Subjectively this feels more confessional, speculative, and probing. I ask the reader to bear with me.

The tentative assertion is that the two-kingdoms doctrine, wherever and whenever it originated, came to serve the purposes of an aristocratic worldview (the "divine right" of kings may be one extremity in this direction) and to hamstring with obstruction and doubt every effort toward democratic self-governance as vulgar meddling in the divinely ordained structures of hierarchy. If this critique is true—I think it is—then we have another example of the way in which aristocratic prerogative, by means of inherited intellectual convention, not only thwarts democratic self-governance but obstructs the kingdom of God as well. I don't mean to strictly equate, as a tight corollary, the kingdom of God with democratic self-governance. I believe the kingdom of God is essentially outside human contrivance or calculated control; it can't be willfully engineered; or, rather, though human control may block and thwart the kingdom of God, human control really cannot call the kingdom of God into being, at least in the sense of conventional engineering rationality. Spirit may yearn to inundate human consciousness with transformative wholeness, but we are not Spirit's self-appointed Army Corps of Engineers.

In simple language, the Gospels tell us to *seek* the kingdom with humility, seek it *first* and all else will be given. The promise seems to be that the kingdom of God will arise and take shape among us precisely as we are significantly humble to permit it to happen and sufficiently loving to allow it to grow. And this, it seems, points a crooked finger at the moon—that

One Last Thing

is, we have to at least earnestly begin to empty ourselves of self-serving willfulness so that we may be guided (perhaps filled) by Spirit.

To those who equate democracy with equality and then go on to ridicule democracy by pointing out, rightly, that human beings are not exactly equal, the Christian response must be that, in any context of survival, in any form, the Christian obligation is to share the loaf, give the second coat, walk the extra mile, and even put the welfare of the other ahead of one's own. So rather than use inequality to build a case for a governing system based on alleged or perceived superiority (which is the etymological essence of "aristocracy"), the Christian is under instruction to be the least, to be a servant, to wash others' dirty feet, even to be deliberately *less* than equal. This "inequality" thus underscores and highlights servanthood and stewardship rather than erasing, blurring, or fudging them as behavioral mandates. But this servanthood is neither obsequiousness nor fawning. It is the true basis for real and deep contact. Whether one starts from a position of equality or inequality is, for the Christian, of relatively minor consequence. The outcome—a disposition toward nonviolent and loving leastness and therefore toward government as servant for all rather than agency in behalf of the rich and powerful—is the same.

This is how the kingdom of God turns aristocracy on its head and folds civilization inside out. The "kingdom" is ethical origami. The kingdom of God requires democracy because decisions need to be made in community, in ongoing, open-ended dialogue that always seeks to be both cleansed of self-seeking and informed by spin-free information. The most comprehensive, overarching term we have for such sincere dialogue is "democracy." In a true democracy, voting is preceded not by immensely expensive and morally repulsive political campaigns that manipulate emotions (especially in the direction of disgust, contempt, and rage), but by a complex mix of exhaustively straightforward discussion, meditation, reflection, and prayer. (That corporate money for political advertisement is currently defended as "free speech," or that the corporations through which money is funneled are considered legal "persons," are tokens of how corrupted our understanding of democracy has become. That money buys a big mouth is obvious. It is the essence of vulgar aristocracy.)

We are taught and told that the kingdom of God can be anywhere and everywhere. This is fine, wonderful, and reassuring. But we also know—because Jesus told us, if we're to believe the Gospels—that the "kingdom" unfolds most fully and gaily where sharing and simplicity are consistently practiced and lived, where humility "rules," where Spirit is truly welcome.

The Kingdom of God Is Green

Democracy cannot endure an aristocracy of wealth and power. Such a "democracy" is a sham. The kingdom of God cannot abide the doctrine of two kingdoms. Otherwise, the Lord's Prayer—Your kingdom come, Your will be done, *on Earth as it is in heaven*—is only meaningless twaddle. As an on-again/off-again news junkie, I think I know what twaddle is. The "news," newspapers, news magazines, and electronic talking heads are fairly well afloat in a toxic bubble bath of twaddle. The clarity and cleanliness of the Gospels is a wonderful respite from such infantile mud puddling.

I suppose I could say that Jesus is my Teacher and my Friend. But if I have misstated, overstated, or even understated anything—any things—in these essays, I ask for forgiveness, correction, and guidance. If any of this is on or even near the mark, I can only say—given the magnitude of irreverence and misbehavior in my life—that I am living proof of Spirit's vast eutopian compassion, and even of—Her?—extremely wry sense of humor.

37

Civilization, 'Civilization', and the Kingdom of God: An Afterword

No doubt there will be readers who, while agreeing with this or that portion of the analysis in these essays, will say my use of "civilization" is much too perverse and way too distorted. I have friends who tell me exactly this and, besides, why don't I find a better, more benign, and less inflammable word to use in its place? Why put a stick in a hornets' nest? When I ask what that less inflammable word might be and why is this a hornets' nest, I get knowing looks but no adequate response. The knowing looks convey an attitude that says let sleeping dogs lie; but everybody knows these dogs are not asleep; they are very much awake, mean, and hungry. To *pretend*, as is our norm, that the dogs are asleep or just cuddly pets is a kind of complicity in denial. This soporific philosophy is not my cup of tea.

So I revert to my best and favorite teacher on the subject, my eutopian gardening master, Lewis Mumford. In "Kings as Prime Movers," Chapter Eight in *The Myth of the Machine*, Mumford closes with a short section entitled "Civilization and 'Civilization.'" I quote it here in its entirety:

> With kingship, power as an abstraction, power as an end in itself, became the chief identifying mark of 'civilization,' as opposed to all earlier norms and forms of culture.
>
> Civilization, still often used as a word of eulogy and admiration, in comparison with what used to be called savagery and barbarism, is taken as a general term to cover law, order,

justice, urbanity, civility, rationality; and it currently implies a cumulative effort to further the arts and sciences and to improve the human condition by continued advances in both technology and responsible government. All these terms of admiration and praise, which seemed in the eighteenth century self-evident and self-justifying, except to an occasional dissident like Rousseau, have now become ironic: at best they represent a hope and a dream that have still to be fulfilled.

Here and hereafter I use the term 'civilization' in quotation marks in a much narrower sense: to denote the group of institutions that first took form under kingship. Its chief features, constant in varying proportions throughout history, are the centralization of political power, the separation of classes, the lifetime division of labor, the mechanization of production, the magnification of military power, the economic exploitation of the weak, and the universal introduction of slavery and forced labor for both industrial and military purposes. These institutions would have completely discredited both the primal myth of divine kingship and the derivative myth of the machine had they not been accompanied by another set of collective traits that deservedly claim admiration: the invention and keeping of the written record, the growth of visual and musical arts, the effort to widen the circle of communication and economic intercourse far beyond the range of any local community: ultimately the purpose to make available to all men the discoveries and inventions and creations, the works of art and thought, the values and purposes that any single group has discovered.

The negative institutions of 'civilization,' which have besmirched and bloodied every page of history, would never have endured so long but for the fact that its positive goods, even though they were arrogated to the use of a dominant minority, were ultimately of service to the whole human community, and tended to produce a universal society of far higher potentialities, by reason of its size and diversity. Even immediately their symbols perhaps attracted those who were only spectators of these achievements. This universal component was present from the beginning, because of the cosmic foundations of royal power: but the efforts to create a universal society were delayed, until our own day, by the lack of adequate technical instruments for rapid transportation and instantaneous communication.

Yet the claim of universality itself, from Naram-Sin to Cyrus, from Alexander to Napoleon, was repeatedly made: one of the last of the 'all-powerful' monarchs, Genghis Khan, proclaimed himself the sole ruler of the entire world. That boast

Civilization, 'Civilization', and the Kingdom of God: An Afterword

was at once an aftermath of the myth of divine kingship and a prelude to the new myth of the machine.[1]

Here, for all my regard and affection for Lewis Mumford, is the clear limitation of what I will call an inadequate secular analysis, that is, an analysis devoid of an explicit spiritual dimension. (In Matthew 4 and Luke 4, what Mumford calls "this universal component" is dealt with pretty starkly when Jesus unconditionally rejects "all this power and the glory of these kingdoms" in the wilderness temptation scene.) In Mumford's passage, the only hope he expresses for civilization in its best and universal form lies in the creation of "adequate technical instruments for rapid transportation and instantaneous communication." To be blunt, I do not believe this is either adequate or true—except insofar as rapid transportation and instantaneous communication are clear and decisive symbols for what Thomas J. J. Altizer calls "new and absolute immanence," that is, symbols in an outward manifestation of what is occurring more profoundly and more intimately in the spiritual and ethical dimension.

Civilization, like capitalism, perhaps like any ideology or "ism," certainly contains contradictions. The most basic of these contradictions is that Creation and everybody (and every thing) in it is a complete and utter gift from the unknown; and that makes Civilization, with its Class and War (to bring Arnold Toynbee's shadow into this afterword), a control throttle, a bottleneck, a chokehold on Creation. If the kingdom of God means anything remotely like what I think it does, and if civilization is at all what I allege—and have, perhaps tiresomely, asserted in these essays—then it's not the absence of "adequate technical instruments" that's responsible for the delay of the universality of civilization in its conventionally understood sense. What has delayed the universality of civilization is 'civilization' itself. A successful thief may become a respected philanthropist, but a thief is still a thief, just as it is easier for a camel to make its way through the eye of a needle than it is for a rich man to enter the kingdom of God. The Gospel represents a kind of Stone Age psychology in that it advocates a fullness of sharing whether there is much or little to share. Civilization, on the other hand, might be willing to share from its excess or its "surplus" if the recipient agrees to certain conditions, including deferential gratefulness, and if certain "structural adjustments" have been made in the economy so as to perpetuate the accumulation of an excess in the hands of a few.

Civilization's chokehold lies not so much in "human nature"— which term is far too much a camouflage for supposedly "unchangeable" behavior, and therefore a term of infinite dismay, despair, sorrow, and

The Kingdom of God Is Green

hopelessness—as it lies in a certain slice of self-justifying male aggression and "democratic" greed. (In Mumford, power "as an abstraction" is identified with kingship, centralization, class separation, division of labor, mechanization, military power, economic exploitation, slavery and forced labor, and evokes such names as Naram-Sin, Cyrus, Alexander, Napoleon, and Genghis Khan. Every image, every tendency, every person is fully and explicitly male.)

Conventional Christianity, especially "conservative" fundamentalism, waits for an angry and judging Father to break into the world and beat the crap out of all the dirty sinners and stubborn unbelievers who've been duly and repeatedly warned. It's really quite easy to see how this imagery of God, of who God is purported to be, correlates exactly to that kind of kingship produced by civilization. It is exactly and precisely the monarchical image or model of God. Civilization is God's cop; God is civilization's Top Cop.

But our dear, disturbing Jesus refuses to fit snugly or comfortably within this neat little self-serving, hierarchical picture. The kingdom of God is all about renouncing control, abandoning power, and defying authority. It's about sharing, healing, being the least, turning the other cheek, and suffering even unto death. The kingdom of God is a spiritual revolution that quietly, persistently, and unrelentingly seeks to revolutionize the totality of human consciousness, conduct, and culture. It does not concentrate on individual or otherworldly "salvation." It does not stop at the door of Congress or at the steps of the White House or even at the Security Council table of the United Nations. It is fully and totally "imperialistic" in that it will stop at nothing less than the leavening of the entire human loaf, including all the human loaf's institutions: only the "imperialism" of this spiritual leaven is entirely benign, radiating political peace and ecological health.

The "God" who breaks into the world comes armed, not with a redhot divine sword and grim legion of wrathful angels itching for a bloody brawl, but with a pan of clean water and a dry towel to bathe our tired, dirty, and distracted minds. We men especially could use a fair amount of this redemptive brainwashing.

Huston Smith, in the 1958 edition of *The Religions of Man*, says in his final chapter that "there is no greater way to depersonalize another than to speak to him without also listening," and that it is "impossible to love another without listening":

Civilization, 'Civilization', and the Kingdom of God: An Afterword

> Those who listen in the present world work for peace, a peace built not upon ecclesiastical or political empire, but upon understanding and the mutual involvement in the lives of others that this brings. For understanding . . . brings respect, and respect prepares the way for a higher power, love—the only power that can quench the flames of fear, suspicion, and prejudice, and provide the means by which the peoples of this great earth can become one to one another.[2]

Perhaps it's time we asked the "bad guys," the "terrorists," what their complaints are, what their beef is. Or do we prefer (in unacknowledged protection of our cherished myths) to keep pretending that they just "hate our freedoms"?

Spirit is neither a Christian nor the Top Gun in a divine SWAT team. Spirit is *love*, but we seem to have such an extremely difficult time saying certain four-letter words, much less putting them into political practice.

Notes

1. Mumford, *Myth*, 186–87.
2. Smith, *Religions*, 355.

A Note on Source Material

THIS PRESENT BOOK, THIS series of essays, does two things in particular. First, it draws conclusions from the previously articulated insights of several recognized thinkers and weighty authors. Second, it also pushes beyond, behind, and beneath the conventionally permissible Gestalt of our collective understanding in an effort to break out of the camp of civilized concentration. The language, at times, is both evasively playful and deliberately confrontational, both "out of the box" and "in your face," to use two expressions with contemporary juiciness.

Allow me to invoke one final example of my "method."

There's a very short (less than a full page) article by Jerry Mander (whose name seems to have been created by the same irrepressible spirit that sometimes inspires my "wit") in the Fall 2001 issue of *Yes!* magazine. Mander is the fairly famous author of a polemic against television. In his *Yes!* article, "Unplug Your Brain," Mander says that television is the "most efficient medium ever invented for cloning corporate consciousness." He talks of our collective electronic addiction as a "weird experiment in mind control":

> Ours is the first generation to have essentially moved its life inside media; to have replaced direct contact with people and nature with simulated, edited, recreated versions. Television is the original "virtual reality."
>
> This situation is really weird. It's almost sci-fi in its feeling and in its possibility for autocratic control—the few speaking to the many.[1]

This observation and analysis is, it seems to me, fundamentally true. Yet our response to it is a kind of mildly embarrassed ho-hum, as we browse for the next or newest electronic gadget, add yet more channels to our cable overload, or scan the satellite heavens for the latest cosmic giggles. We know it's true that our collective addiction to electronic "culture" has been both cause and effect of our loss of local culture, just as we

A Note on Source Material

know that agribusiness in full vertical integration has wrecked agrarian culture, as consolidated school systems have numbed and demoralized local communities. We all know what happens to small, local businesses when Wal-Mart moves in. We know this. But neither we nor those we elect to public office do much about it. In fact, we do almost nothing about it. If we try to do something about it, we will have the flag flapped in our faces: we are transparently against "freedom," against "free enterprise," against "the American way," and maybe even against "God." That is, the civilized system—including formal governance (with its three branches), the economy (with its "free market" ideology), the education system (with its drive for "professional" advancement), the religious institutions (with their specialized focus on "salvation"), etc.—has accrued a depth and magnitude of conditioning power far beyond anything achieved by ancient empire. Never before has the entire world been brought so closely under the pervasive electronic influence of civilized mind control. Never before has local culture been made to feel—Mander's words—so "backward" and "unworthy."[2]

From a rural perspective I have seen this process at work since my youth. I'm just old enough—born in 1946—to have witnessed a briefly blossoming local agrarian culture poisoned and starved by the political and economic undermining of it's small-scale agrarian way of life, the forced removal of its schools, the withdrawal of its church, and its fascinated, ambiguously voluntary submission to electronic "culture." And as local culture withered and died, the "brand names" of civilization infiltrated and claimed allegiance.

The process is, it seems to me, indisputable. The question that hangs out there is—So what? Why should anybody care? Isn't it a case of Good Riddance? The only good Indian is a dead Indian. We had to destroy the village in order to save it. Progress is our most important product. Better dead than Red.

One need not idealize local culture in order to recognize its ecological value. Conversely, one need not stand with the civilized SWAT team, gloating over the genocide of villages, in order to recognize certain serious constraints within village culture, perhaps its gender configurations predominantly.

The problem, or at least a huge aspect of the problem, is that "civilized values" is an empowered concept of totalitarian magnitude. Civilization is to local culture, as John Dominic Crossan has told us, what the classical aristocracy was to the peasantry: somewhere beyond, in another world. It is nearly impossible to find social acceptability or cultural credibility outside of "civilization." Anything "uncivilized" is already on the UnAmerican Activities list—scapegoated and blacklisted without so much as a political peep. We no longer have—for the first time in human history—a noncivilized culture with explicit folk roots and self-provisioning foundations.

The *system* therefore has the moral high ground, the bully pulpit, the institutions, the weapons and satellites, the security contractors, and it has an explanatory ideology that can rationalize and blend all kinds of political sausage. Spin doctors can grind you a hot dog on global warming, rogue states, or free trade in a matter of moments. (Would you like a beverage with your climate change, Sir?) But what the system doesn't have is soul, Spirit, or a true sense of the eternal.

Whatever the raggedness of the antiglobalization crew (or its current "Occupy Wall Street" incarnation), whatever the wrongheadedness of this blithe assertion or that rash action, the "anti"globalization folks really *do* stand for a world of stewardship and sharing. They mean it. Against all the impossible odds of the civilized juggernaut, they have soul, Spirit, and the eternal on their side. Even more than that: soul, Spirit, and the eternal are notorious saboteurs, secret agents, undercover fifth columnists, who infiltrate the ranks of the civilized "enemy," transforming consciousness in devious ways and reconstructing both political outlook and spiritual conviction.

One wishes them, with all the etymological quirks in *Webster's*, Godspeed—including the never obsolete usages of "in the nick of time" and "in conclusion."

Notes

1. Mander, "Unplug," http://www.yesmagazine.org/issues/technology-who-chooses/460.

2. Mander, "Unplug," http://www.yesmagazine.org/issues/technology-who-chooses/460.

Bibliography

Abbey, Edward. *Desert Solitaire*. New York: Ballantine, 1985.
Altizer, Thomas J. J. *The Gospel of Christian Atheism*. Philadelphia: Westminster Press, 1966.
Araghi, Farshad. "The Great Global Enclosure of our Times: Peasants and the Agrarian Question at the End of the Twentieth Century." In *Hungry for Profit: The Agribusiness Threat to Farmers, Food, and the Environment*, edited by Fred Magdoff, John Bellamy Foster, and Frederick H. Buttel. New York: Monthly Review Press, 2000.
Armitage, Angus. *The World of Copernicus*. New York: Mentor Books, 1951.
Augustinas, Aurelius. *The City of God*, translated by Marcus Dods. New York: Random House, 1950.
Bailie, Gil. *Violence Unveiled: Humanity at the Crossroads*. New York: The Crossroad Publishing Company, 1995.
Bainton, Roland. *Here I Stand: A Life of Martin Luther*. Nashville: Abingdon, 1950.
Barsamian, David. "The Progressive Interview." In *The Progressive*, September 2002.
———. "The Progressive Interview." In *The Progressive*, May 2004.
Berrigan, Phil. "Pray Always and Never Lose Hope." In *Catholic Agitator*, February 2002.
Berry, Wendell. "Back to the Land: The Radical Case for Local Economy." In *The Amicus Journal*, Winter 1999.
Block, Fred. "The Right's Moral Trouble." In *The Nation*, September 30, 2002.
Bonhoeffer, Dietrich. *Letters and Papers from Prison*, edited by Eberhard Bethge. New York: Macmillian, 1979.
Borg, Marcus. *God at 2000*. Harrisburg, PA: Morehouse Publishing, 2000.
———. *Meeting Jesus Again for the First Time*. San Francisco: Harpers, 1994.
———. *The God We Never Knew*. San Francisco, Harpers, 1997.
———. "The Palestinian Background for a Life of Jesus." In *The Search for Jesus: Modern Scholarship Looks at the Gospels*, edited by Hershel Shanks. Washington, D.C.: Biblical Archaeology Society, 1994.
Boulding, Elise. *The Underside of History*. Boulder, CO: Westview Press, 1976.
Brown, Norman O. *Life Against Death: The Psychoanalytical Meaning of History*. Middletown: Wesleyan University Press, 1959.
———. *Love's Body*. New York: Vintage Books, 1966.
Brownfeld, Allan. "Strange Bedfellows: The Jewish Establishment and the Christian Right." In *Washington Report on Middle East Affairs*, August 2002.
———. "Will Election Year Democratic Politics Derail 'Road Map' to Mideast Peace?" In *Washington Report on Middle East Affairs*, September 2003.
Buber, Martin. *Paths in Utopia*. London: Routledge & Kegan Paul, 1949.

Bibliography

Buchanan, George Wesley. "Misunderstandings About Jerusalem's Temple Mount." In *Washington Report on Middle East Affairs*, August 2011.
Buchanan, Patrick. *The Death of the West*. New York: St. Martin's Press, 2002.
Bunch, Charlotte. "Whose Security?" In *The Nation*, September 23, 2002.
Caro, Robert A. *Means of Ascent*. New York: Alfred A. Knopf, 1990.
———. *Master of the Senate*. New York: Alfred A. Knopf, 2002.
Carroll, James. *Constantine's Sword: The Church and the Jews*. Boston: Houghton Mifflin, 2001.
Cayley, David. The *Corruption of Christianity: Ivan Illich on Gospel, Church and Society*. Toronto: CBC Ideas Transcripts, PO Box 500, Station A, Toronto, ON n5w 1e6, 2000.
———. *Ivan Illich in Conversation*. Concord, Ontario: House of Anansi Press, 1992.
Childe, Gordon. *What Happened in History*. Baltimore: Penguin Press, 1954.
Chittister, Joan. "God Become Infinitely Larger." In *God at 2000*, edited by Marcus Borg and Ross Mackenzie. Harrisburg, PA: Morehouse Publishing, 2000.
———. *Wisdom Distilled from the Daily: Living the Rule of St. Benedict Today*. New York: HarperCollins, 1990.
Chomsky, Noam. "The Journalist from Mars." In *Extra!*, April 2002.
———. "The Progressive Interview." In *The Progressive*, May 2004.
———. *9-11*. New York: Seven Stories Press, 2001.
Cockburn, Alexander. "So Who's the Fascist Here?" In *The Nation*, May 21, 2012.
Cook, Christopher D. "A Progressive Interview with Naomi Klein." In *The Progressive*, December 2011/January 2012.
Cornwell, John. *Hitler's Pope: The Secret History of Pius XII*. New York: Viking, 1999.
Crossan, John Dominic. *The Historical Jesus: The Life of a Mediterranean Jewish Peasant*. San Francisco: HarperSanFrancisco, 1991.
———. "The Passion, Crucifixion and Resurrection." In. *The Search for Jesus: Modern Scholarship Looks at the Gospel*, edited by Hershel Shanks. Washington, D.C.: Biblical Archaeology Society, 1994.
———. "The Infancy and Youth of the Messiah." In *The Search for Jesus: Modern Scholarship Looks at the Gospel*, edited by Hershel Shanks. Washington, D.C.: Biblical Archaeology Society, 1994.
———. *Jesus: A Revolutionary Biography*. San Francisco: HarperSanFrancisco, 1995.
Daly, Herman. "Economics and Sustainability: In Defense of a Steady-State Economy." In *Deep Ecology*, edited by Michael Tobias. San Diego: Accent Books, 1985.
Daly, Mary. *Gyn/Ecology: The Metaethics of Radical Feminism*. Boston: Beacon Press, 1978.
de Riencourt, Amaury. *The Coming Caesars*. New York: Coward-McCann, Inc., 1957.
De Rosa, Peter. *Vicars of Christ: The Dark Side of the Papacy*. New York: Crown Publishers, Inc., 1988.
Dietrich, Jeff. "Pray Always and Never Lose Hope." In *Catholic Agitator*, February 2002.
———. "Seminary in the Cellblock." In *The Other Side*, March & April 2002.
Douglas, William O. *Russian Journey*. Garden City: Doubleday & Company, 1956.
Douglas-Klotz, Neil. "Original Prayer." In *Sounds True*, Summer 2000.
Dyer, Joel. *Harvest of Rage: Why Oklahoma City Is Only the Beginning*. Boulder, CO: Westview Press, 1997.
Ehrenreich, Barbara. *Blood Rites: Origins and History of the Passions of War*. New York: Metropolitan Books, 1991.

Bibliography

Ellsberg, Robert. *All Saints*. New York: Crossroad Publishing, 1997.
Encyclopaedia Britannica. 24 volumes. London, 1936.
Engel, J. Ronald. "Liberal Democracy and the Fate of the Earth." In *Spirit and Nature: Why the Environment Is a Religious Issue*, edited by Steven C. Rockefeller and John C. Elder. Boston: Beacon Press, 1992.
Gilk, Paul. *Green Politics Is Eutopian*. Eugene, OR: Wipf and Stock, 2008.
———. *Nature's Unruly Mob: Farming and the Crisis in Rural Culture*. Eugene, OR: Wipf and Stock, 2009.
———. *Polemics and Provocations: Essays in Anticipation of the Daughter*. Eugene, OR: Wipf and Stock, 2010.
Gish, Arthur G. *The New Left and Christian Radicalism*. Grand Rapids: Wm. B. Eerdmans, 1970.
Gonzalez, Justo. *The Story of Christianity: The Reformation to the Present Day*. New York: Harper Collins, 1985.
Goodman, Paul. *Compulsory Miseducation*. New York: Horizon, 1964.
Goodwyn, Lawrence. *The Populist Moment: A Short History of the Agrarian Revolt*. New York: Oxford University Press, 1978.
Gorenberg, Gershom. *The End of Days: Fundamentalism and the Struggle for the Temple Mount*. New York: The Free Press, 2000.
Harrington, Michael. *Socialism*. New York: Dutton, 1992.
Hawken, Paul. "The Resurgence of Citizens' Movements." In *Utne Reader*, November-December, 2000: http://www.utne.com/200011-01//TheResurgenceofCitizensMovements.aspx.
Hedges, Chris. *War is a Force that Gives Us Meaning*. New York: Public Affairs, 2002.
Heilbroner, Robert. *An Inquiry into the Human Prospect*. New York: W. W. Norton & Company, 1974.
Heyward, Carter. *Our Passion for Justice: Images of Power, Sexuality, and Liberation*. Cleveland: The Pilgrim Press, 1984.
Hochschild, Adam. *King Leopold's Ghost*. New York: Houghton Mifflin, 1998.
Illich, Ivan. *Deschooling Society*. New York: Harper & Row, 1971.
Jerusalem Bible. Garden City: Doubleday, 1966.
Johnson, Warren. *Muddling Toward Frugality*. Boulder, CO: Shambhala, 1978.
Johnston, David Cay. *Perfectly Legal*. New York: Penguin, 2003.
Jung, Carl. *Memories, Dreams, Reflections*. Recorded and edited by Aniela Jaffé. Translated from the German by Richard and Clara Winston. New York: Vintage Books, 1965.
Kaplan, Robert. "Looking the World in the Eye." In *The Atlantic Monthly*, December 2001.
Klein, Naomi. "Capitalism vs. the Climate." In *The Nation*, November 28, 2011.
———. *The Shock Doctrine: The Rise of Disaster Capitalism*. New York: Metropolitan Books, 2007.
Klinker, Philip. "The Base Camp of Christendom." In *The Nation*, March 11, 2002.
Kohr, Leopold. *The Breakdown of Nations*. New York: E. P. Dutton, 1978.
Krebs, A. V. *The Corporate Reapers: The Book of Agribusiness*. Washington, D.C.: Essential Books, 1992.
Kuhns, William. *In Pursuit of Dietrich Bonhoeffer*. Garden City, NJ: Image Books, 1969.
Lappé, Anna and Frances Moore Lappé. *Hope's Edge*. New York: Putnam, 2002.
Lewis, C. S. *That Hideous Strength*. New York: Macmillan, 1968.

Bibliography

Lind, Michael. *Made in Texas: George W. Bush and the Southern Takeover of American Politics.* New York: Basic Books, 2003.
Mander, Jerry. "Unplug Your Brain." In *Yes!*, Fall 2001.
McCarthy, Colman. *Solutions to Violence.* Washington, D.C.: Center for Teaching Peace, [no date].
McNeill, William H. *The Rise of the West: A History of the Human Community.* New York: Mentor Books, 1965.
Merton, Thomas. *Faith and Violence.* Notre Dame: University of Notre Dame, 1968.
Mitchell, Stephen. *The Gospel According to Jesus.* New York: HarperCollins, 1991.
Moyers, Bill. "Our Story." In *The Progressive*, May 2004.
Mumford, Lewis. *The City in History.* New York: Harcourt, Brace, and Jovanovich, 1961.
———. *Interpretations and Forecasts: 1922–1972.* New York: Harcourt, Brace, and Jovanovich, 1973.
———. *The Myth of the Machine.* New York: Harcourt, Brace, and World, 1966.
———. *The Pentagon of Power.* New York: Harcourt, Brace, and Jovanovich, 1970.
———. *The Transformations of Man.* New York: Harper Row, 1972.
Myers, Ched. *Binding the Strong Men: A Political Reading of Mark's Story of Jesus.* Maryknoll, NY: Orbis, 2008.
Nader, Ralph. *Crashing the Party.* New York: St. Martin's Press, 2002.
Newcomb, Steven. "The Legacy of Religious Racism in U.S. Indian Law." In *Indian Country Today*, April 24, 2002.
Nichols, John. "Dark Ages Ahead at the NLRB." In *The Nation*, September 3/11, 2001.
Niebuhr, Reinhold. *Faith and Politics: A Commentary on Religious, Social and Political Thought in a Technological Age*, edited by Ronald H. Stone. New York: George Braziller, 1968.
Nolan, Albert. *Jesus Before Christianity.* Maryknoll, NY: Orbis Books, 2001.
Norberg-Hodge, Helena. *Ancient Futures: Learning from Ladahk.* Oxford: Oxford University Press, 1992.
O'Brien, Tim. *The Things They Carried.* New York: Broadway Books, 1990.
Parenti, Christian. "Ideology or Electricity." In *The Nation*, May 7, 2012.
———. *Tropic of Chaos: Climate Change and the New Geography of Violence.* New York: Nation Books, 2011.
Parrington, Vernon Lewis. *The Colonial Mind: 1620–1800.* New York: Harcourt, Brace & World, 1927.
Phillips, Kevin. "DYNASTIES! How their wealth and power threaten democracy." In *The Nation*, July 8, 2002.
———. *Wealth and Democracy.* New York: Broadway Books, 2002.
Pollitt, Katha. "Invisible Women." In *The Nation*, April 4, 2005.
Randall, Steve. "Take No Prisoners." In *Extra!*, April 2002.
Roy, Arundhati. "Fascism's Firm Footprint in India." In *The Nation*, September 30, 2002.
Rudin, A. James. "Centuries of 'Jew Hatred' Brought to Light." In *National Catholic Reporter*, February 2, 2001.
Schumacher, E. F. *Small is Beautiful: Economics as if People Mattered.* New York: Harper & Row, 1973.
Sergio, Lisa. *Jesus and Woman.* McLean, VA: EPM Publications, Inc., 1975.
Shiva, Vandana. *Staying Alive: Women, Ecology and Development.* Brooklyn: South End Press, 2010.

Slattery, W. Michael. *Jesus the Warrior? Historical Christian Perspectives and Problems on the Morality of War and the Waging of Peace.* Milwaukee: Marquette University Press, 2007.
Smith, Henry Nash. *Virgin Land: Myth and Symbol in the American West.* New York: Vintage Books, 1950.
Smith, Huston. *The Religions of Man.* New York: Harper & Row, 1958.
Spong, John Shelby. *A New Christianity for a New World.* San Francisco: HarperSanFrancisco, 2001.
Stillman, Deanne. "Onward, Christian Soldiers." In *The Nation*, June 3, 2002.
Stringfellow, William. *A Keeper of the Word.* Grand Rapids: Wm. B. Eerdmans, 1994.
Sullivan, Andrew. "Christianity's Original Sin." In *The New York Times Book Review*, January 14, 2001.
Temple, Katharine. "Lessons from Nuremberg." In *The Catholic Worker*, March-April 2002.
Thompson, E. P. *The Making of the English Working Class.* New York: Vintage Books, 1963.
Thompson, William Irwin. *The Time Falling Bodies Take to Light: Mythology, Sexuality, and the Origins of Culture.* New York: St. Martin's Press, 1981.
Toynbee, Arnold J. *Civilization on Trial.* New York: Oxford University Press, 1948.
Vidal, Gore. *Perpetual War for Perpetual Peace: How We Got To Be So Hated.* New York: Thunder's Mouth Press/Nation Books, 2002.
———. *The Smithsonian Institution.* New York: Random House, 1998.
Webster, David Kenyon. *Parachute Infantry.* New York: Dell, 1994.
Wills, Garry. *Papal Sin: Structures of Deceit.* New York: Doubleday, 2000.
Wood, Ellen Meiksins. "The Agrarian Origins of Capitalism." In *Hungry for Profit: The Agribusiness Threat to Farmers, Food, and the Environment,* edited by Fred Magdoff, John Bellamy Foster, and Frederick H. Buttel. New York: Monthly Review Press, 2000.
Zinn, Howard. *A People's History of the United States.* New York: Harper Colophon Books, 1980.
———. *Terrorism and War.* New York: Seven Stories Press, 2002.

www.ingramcontent.com/pod-product-compliance
Lightning Source LLC
Chambersburg PA
CBHW070304230426
43664CB00014B/2632